THE LITTLE GOLD GRAMMAR BOOK

MASTERING THE RULES
THAT UNLOCK THE POWER
OF WRITING

Brandon Royal

Published by Maven Publishing

MAVEN

© 2010 by Brandon Royal

All rights reserved. No part of this work may be reproduced or transmitted in any form or by any means, electronic or mechanical—including photocopying, recording, or any information storage and retrieval system—without permission in writing from the author or publisher.

Published by:

Maven Publishing
4520 Manilla Road
Calgary, Alberta, Canada T2G 4B7
www.mavenpublishing.com

Library and Archives Canada Cataloguing in Publication:

Royal, Brandon
The little gold grammar book : mastering the rules that unlock the power of writing / by Brandon Royal.

ISBN 978-1-897393-30-7

Library of Congress Control Number: 2009909354

In addition to the paperback edition, this book is available as an eBook and in the Adobe PDF file format.

Technical Credits:
Cover Design: George Foster, Fairfield, Iowa, USA
Editing: Jonathan K Cohen, Irvine, California, USA

This book's cover text was set in Minion. The interior text was set in Scala and Scala Sans.

Contents

Introduction		5
Chapter 1:	**The 100-Question Quiz**	**7**
	Subject-Verb Agreement	8
	Pronoun Usage	13
	Modification	17
	Parallelism	19
	Comparisons	22
	Verb Tenses	24
	Diction Review	28
	Idioms Review	30
	Answers to the 100-Question Quiz	34
Chapter 2:	**Grammatical Munchkins**	**55**
	The Eight Parts of Speech	56
	Parts of Speech vs. The Seven Characteristics	58
	Other Grammatical Terms	61
Chapter 3:	**Word Gremlins**	**73**
	Diction Showdown	74
	200 Common Grammatical Idioms	94
Chapter 4:	**Putting It All Together**	**101**
	30 All-Star Grammar Problems	102
	Answers and Explanations	121
Editing I – Tune-up		149
Editing II – Punctuation Highlights		175
American English vs. British English		195
Traditional Writing vs. Digital Writing		203
Selected Bibliography		209
Index		211
About the Author		

Introduction

This book is based on a simple but powerful observation: Individuals who develop outstanding grammar skills do so primarily by mastering a limited number of the most important grammar rules, which they use over and over. What are these recurring rules? The answer to this question is the basis of this book.

In everyday writing, we may not always think it necessary to know the intimate rules of grammar. As long as we reach our goal of communicating through our writing, why is there need to know precisely how everything works? After all, as long as we can drive a car, why do we need to know how its engine works?

The individual who desires to become truly proficient in basic writing skills needs to look under the "hood" to better understand how the writing engine works. How do we navigate this magnificent ensemble—the act of accurately stringing words and word groups into sentences? The answer is part science and part art. First, we review the common categories that capture the vast majority of recurring grammar problems. Next, we familiarize ourselves with the names of grammatical terms that describe the words and word groups used in building sentences. Lastly, we see how these parts interact as a whole by testing them through multiple-choice problems that integrate key concepts.

Our journey would not be complete without coverage of a few areas that closely impact "grammar." These include diction and idioms, editing tips, American English versus British English, and traditional writing versus digital writing.

This book is your "grammar mechanic's" guide to writing.

Let's get started.

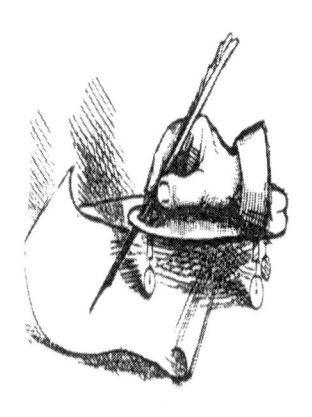

Chapter 1
The 100-Question Quiz

The following 100-question quiz (Q1 to Q100) provides a highly distilled review of grammar, diction, and idioms. The first segment of this quiz addresses grammar and is built on the "big six" grammar categories: subject-verb agreement, modification, pronoun usage, parallelism, comparisons, and verb tenses. The "big six" grammar categories provide a way to break grammar into those areas where errors are most likely to occur. Once we study the rules within each category, we can immediately apply them to many practical writing situations.

Answers to the 100-question quiz are found on pages 33–52. Many of the terms used in chapter 1 are defined in chapter 2 and chapter 3, and these chapters can be reviewed first, before attempting quiz questions, if a more technical grounding is desired. Special notes, marked by (*NOTE* ☙), provide additional commentary when applicable.

Subject-Verb Agreement

The overarching principle regarding subject-verb agreement is that singular subjects require singular verbs while plural subjects take plural verbs. Our objective is to identify the subject in order to determine whether the verb is singular or plural.

☞ Rule: **"And" always creates a compound subject.**

Q1 An office clerk and a machinist (was / were) present but unhurt by the on-site explosion.

The only connecting word that can make a series of singular nouns into a plural subject is "and." In fact, "and" always creates a plural subject with but one exception, as noted in the next rule.

☞ Rule: **If two items joined by "and" are deemed to be a single unit, then the subject is considered singular, and a singular verb is required.**

Q2 Eggs and bacon (is / are) Tiffany's favorite breakfast.

☞ Rule: **"Pseudo-compound subjects" do not make singular subjects plural.**

Pseudo-compound subjects include the following: *as well as, along with, besides, in addition to,* and *together with.*

Q3 A seventeenth-century oil painting, along with several antique vases, (has / have) been placed on the auction block.

☞ Rule: Prepositional phrases (i.e., phrases introduced by a preposition) can never contain the subject of a sentence.

Some of the most common prepositions include *of, in, to, by, for,* and *from.* A definition of the word "preposition," as well as a glossary of other grammatical terms, can be found in chapter 2.

Q4   The purpose of the executive, administrative, and legislative branches of government (is / are) to provide a system of checks and balances.

☞ Rule: **"There is/there are" and "here is/here are" constructions represent special situations where the verb comes before the subject, not after the subject.**

The normal order in English sentences is subject-verb-object (think S-V-O). "There is/there are" and "here is/here are" sentences are tricky because they create situations in which the verb comes before the subject. Thus, these sentence constructions require that we look past the verb—"is" or "are" in this case—in order to identify the subject.

Q5   Here (is / are) the introduction and chapters one through five.

Q6   (Is / are) there any squash courts available?

NOTE ∞ It is a common mistake, especially when writing emails, to use the singular contraction "here's" in referring to a plural subject. Consider this sentence: "Here's the pictures you asked about." The contraction "here's" stands for "here is." The sentence thereby reads: "Here is the pictures you asked about." To correct this, we should avoid the contraction "here's" and write "Here are the pictures you asked about." On the other hand, it would be correct to write: "Here's the list you were looking for." The singular "list" matches the verb "is."

☞ Rule: When acting as subjects of a sentence, gerunds and infinitives are always singular and require singular verbs.

Q7 Entertaining multiple goals (makes / make) a person's life stressful.

Q8 To plan road trips to three different cities (involves / involve) the handling of many details.

Exhibit 1.1 Chart of Indefinite Pronouns

Singular or Plural	Examples
Certain indefinite pronouns are always singular	anybody, anyone, anything, each, either, every, everybody, everyone, everything, neither, nobody, no one, nothing, one, somebody, someone, something
Certain indefinite pronouns are always plural	both, few, many, several
Certain indefinite pronouns can be either singular or plural	all, any, most, none, some

☞ Rule: "-One," "-body," and "-thing" indefinite pronouns are always singular.

Q9 One in every three new businesses (fails / fail) within the first five years of operation.

☞ Rule: Certain indefinite pronouns—"both," "few," "many," and "several"—are always plural.

Q10 Few of the students, if any, (is / are) ready for the test.

☞ Rule: "Some" and "none" indefinite pronouns may be singular or plural.

Q11 Some of the story (makes / make) sense.

Q12 Some of the comedians (was / were) hilarious.

Q13 None of the candidates (has / have) any previous political experience.

☞ Rule: In "either...or" and "neither...nor" constructions, the verb matches the subject which comes directly after the "or" or "nor."

Q14 Either Johann or Cecilia (is / are) qualified to act as manager.

Q15 Neither management nor workers (is / are) satisfied with the new contract.

☞ Rule:	Collective nouns denote a group of individuals (e.g., family, government, assembly, crew). If the collective noun refers to a group as a whole or the idea of oneness predominates, use a singular verb. If not, use a plural verb.
Q16	Our group (is / are) meeting at 6 p.m.
Q17	A group of latecomers (was / were) escorted to their seats.
☞ Rule:	"The number" is a singular noun and takes a singular verb. "A number" is plural and takes a plural verb.
Q18	The number of road accidents (has / have) decreased.
Q19	A number of train accidents (has / have) occurred.
☞ Rule:	Percents or fractions, when followed by an "of phrase," can take a singular or plural verb. The key lies in determining whether the noun within the "of phrase" is singular or plural.
Q20	Fifty percent of video gaming (is / are) having great reflexes.
Q21	Two-thirds of their classmates (has / have) wakeboards.
☞ Rule:	Measurements involving money (e.g., dollars, pounds), time (e.g., five years, the fifties), weight (e.g., pounds, kilograms), or volume

(e.g., gallons, kilograms) are always singular and take singular verbs.

Q22   Ten dollars (is / are) an average daily wage for many people in the developing world.

Pronoun Usage

Problems relating to pronoun usage typically center on personal pronouns. Three areas of confusion may include: choosing between the subjective or objective forms of personal pronouns, making sure pronouns agree in number with their antecedents, and ensuring that pronouns are not ambiguous in context.

EXHIBIT 1.2   CHART OF PERSONAL PRONOUNS

	Subjective	Possessive	Objective
first-person singular	I	my, mine	me
second-person singular	you	your, yours	you
third-person singular	he, she, it	his, her, hers, its	him, her, it
first-person plural	we	our, ours	us
second-person plural	you	your, yours	you
third-person plural	they	their, theirs	them
who	who	whose	whom

☞ Rule: As a general guide, pronouns at or near the front of a sentence take their subjective forms; pronouns at or near the back of a sentence take their objective forms. The precise rule, however, is that pronouns take their subjective form when they are subjects of a verb; they take their objective form when they are objects of a verb.

Q23 The present is from Beth and (she / her).

Q24 Cousin Vinny and (he / him) are both valedictorians.

☞ Rule: Pronouns take their objective form when they are the direct objects of prepositions.

Q25 Between you and (I / me), this plan makes a lot of sense.

Q26 Do not ask for (who / whom) the bell tolls.

Q27 People like you and (I / me) should know better.

☞ Rule: When forming comparisons using "than" or "as ... as," supply any "missing words" (e.g., a verb in the examples below) in order to determine whether the subjective or objective form of the pronoun is correct.

Q28 My son is more sports minded than (I / me).

Q29 We skate as fast as (they / them).

Q30 During our group presentation, our teacher asked you more questions than (I / me).

☞ Rule:	Who vs. Whom. "Who" is the subjective form of the pronoun, and "whom" is the objective form of the pronoun. If "he," "she," or "they" can be substituted for a pronoun in context, the correct form is "who." If "him," "her," or "them" can be substituted for a pronoun in context, the correct form is "whom."
Q31	The woman (who / whom) is responsible for pension planning is Mrs. Green.
Q32	This gift is intended for (who / whom)?
☞ Rule:	Do not use a reflexive pronoun (a pronoun ending in "-self") if an ordinary personal pronoun will suffice.
Q33	The tour leader told Julie and (me / myself) to turn off our cell phones.
Q34	Young Robert hurt (him / himself) while climbing alone.
☞ Rule:	Pronouns must agree in number with their antecedents.
Q35	A not-for-profit, like any other organization, has (its / their) own rules and regulations to follow.
Q36	Everybody should mind (his or her / their) own business.
	NOTE ☙ There is something known today as the "singular they." Although it is not considered proper in formal writing (and formal speech), in informal writing (and colloquial speech), it

is ever common to see or hear the word "they" used to refer to a singular subject. For example: "A parent knows that they have to be involved in a child's education." Although "parent" is singular, it is matched with the plural pronoun "they."

☞ Rule: Pronouns should not be ambiguous in context. If a pronoun does not refer clearly to a specific noun, it results in a situation of "ambiguous pronoun reference."

Ambiguous Sam never argues with his father when <u>he</u> is drunk.

Q37 Sam never argues with his father when _____ is drunk.

☞ Rule: "Pronoun shifts," also known as "shifts in point of view," involve the inconsistent matching of pronouns, either in terms of person or number. Within a single sentence (and perhaps within an entire paragraph or writing piece), first person should be matched with first person, second person matched with second person, and third person matched with third person. A common violation involves matching the third-person "one" or "a person" with the second-person "you." Another violation involves matching the third-person singular "he," "she," "one," or "a person" with the third-person plural "they."

Incorrect To know that <u>a person</u> can't vote is to know that <u>you don't</u> have a voice.

Q 38   To know that a person can't vote is to know that _____ have a voice.

Incorrect   <u>One</u> cannot really understand another country until <u>they</u> have studied its history and culture.

Q 39   One cannot really understand another country until _____ studied its history and culture.

Modification

Modifiers, including modifying phrases, must be placed as close as possible to the nouns they modify. As a mostly uninflected language, English depends heavily on word order to establish modifying relationships. Therefore, the position of words is important. Confusion occurs because most modifiers attach themselves to the first thing they can "get their hands on" in the sentence, even if it isn't the right thing.

☞ Rule: **A misplaced modifier refers to a word which, because of its placement within a sentence, no longer modifies what it originally was intended to modify.**

Incorrect   He told her he wanted to marry her frequently.

Q 40   He _____ told her he wanted to marry her.

Incorrect   Coming out of the wood, the janitor was surprised to find termites.

Q 41   The janitor was surprised to find termites _____ _____.

☞ Rule: **A dangling modifier refers to a situation in which the thing being modified is absent from the sentence.**

Incorrect After writing the introduction, the rest of the report was easy.

Q42 After writing the introduction, _____ easily drafted the rest of the report.

Incorrect Walking along the shore, fish could be seen jumping in the lake.

Q43 Walking along the shore, _____ could see fish jumping in the lake.

☞ Rule: **Occasionally, a modifier or modifying phrase may accidentally be placed where it could modify either of the two words or phrases. This situation results in a "squinting modifier." Because it is unclear which of two words or phrases are being modified, the writer should consider rewriting this sentence to clear up this ambiguity.**

Incorrect She said in her office she had a copy of the map.

Q44 She said she had _____ lying in her office.

☞ Rule: **Whenever a sentence opens with a phrase or clause that is set off by a comma, check to make sure that the first word that follows the comma is properly being modified by the opening phrase or clause that precedes it.**

Incorrect	In addition to building organizational skills, the summer internship also helped me hone my team-building skills.
Q45	In addition to building organizational skills, _____.
Incorrect	An incredibly complex mechanism, there are some 10 billion nerve cells in the brain.
Q46	An incredibly complex mechanism, _____ has some 10 billion nerve cells.
Incorrect	Based on our observations, the project will succeed.
Q47	_____.

Parallelism

Parallelism is both a style issue and a grammar issue. In other words, certain elements of parallelism are based on principle and are deemed to be more effective or less effective, better or worse, while other elements are based on rules and are considered correct or incorrect, right or wrong.

The overarching principle regarding parallelism is that similar elements in a sentence must be written in similar form.

☞ Rule:	**Verbs should follow consistent form. Typically this means that all verbs should end in "-ed" or "-ing."**
Incorrect	In the summer before college, Max <u>was</u> a waiter at a restaurant, <u>pursued</u> magazine sales, and even had a stint at <u>delivering</u> pizzas.

Q48   In the summer before college, Max _____ tables, _____ magazines, and even _____ pizzas.

☞ Rule: When prepositions are used before items in a series of three, there are two possibilities with regard to their use. Either a single preposition is used before the first item in a series (but not with the next two items) or prepositions are used before each item in the series.

Incorrect   Our neighbors went to London, Athens, and to Rome.

Q49   Our neighbors went _____ London, Athens, and Rome.

Q50   Our neighbors went _____ London, _____ Athens, and _____ Rome.

☞ Rule: Correlative conjunctions (i.e., either ... or, neither ... nor, not only ... but also, both ... and, whether ... or, and just as ... so too) require that parallelism be maintained after each component part of the correlative.

Incorrect   Jonathan not only likes rugby but also kayaking.

Q51   Jonathan _____ rugby but also kayaking.

Q52   Jonathan _____ rugby but also _____ kayaking.

THE 100-QUESTION QUIZ

☞ Rule: Gerunds and infinitives should be presented in parallel form. Where possible, gerunds are matched with gerunds and infinitives are matched with infinitives.

Less effective <u>Examining</u> the works of William Shakespeare—his plays and poetry—is <u>to marvel</u> at one man's seemingly incomparable depth of literary expression.

Q53 _____ the works of William Shakespeare—his plays and poetry—is <u>to marvel</u> at one man's seemingly incomparable depth of literary expression.

☞ Rule: At times we can acceptably omit words in a sentence and still retain clear meaning. To check for faulty parallelism (in this context also known as improper use of ellipsis), complete each sentence component and make sure that each part of the sentence can stand on its own.

Incorrect In the *Phantom of the Opera* play, the story <u>is</u> intriguing and the singers superb.

Q54 In the *Phantom of the Opera* play, the story <u>is</u> intriguing and the singers _____ superb.

Incorrect The defendant's own testimony on the stand neither contributed nor detracted from his claim of innocence.

Q55 The defendant's own testimony on the stand neither contributed _____ nor detracted from his claim of innocence.

Comparisons

The overarching principle in comparisons requires that we compare apples with apples and oranges with oranges.

☞ Rule: The superlative ("-est") is used when comparing three or more persons or things; the comparative ("-er") is used when comparing exactly two persons or things.

Q56 Between Tom and Brenda, Tom is (better / best) at math.

Q57 Among our group, Jeff is the (wealthier / wealthiest) person.

Q58 Of all the roses in our neighborhood, Chauncey Gardiner's grow the (more / most) vigorously.

Q59 Chauncey Gardiner's roses grow (more / most) vigorously than any other in the neighborhood.

☞ Rule: Remember to compare the characteristics of one thing to the characteristics of another thing, not the characteristics of one thing directly to another thing.

Incorrect Tokyo's population is greater than Beijing.

Q60 Tokyo's population is greater than the _____ of Beijing.

Q61 Tokyo's population is greater than Beijing's _____.

Q62	Tokyo's population is greater than that of _____.
Q63	Tokyo's population is greater than _____.
Incorrect	Of all the countries contiguous to India, Pakistan's borders are most strongly defended.
Q64	Of all the countries contiguous to India, _____ _____.
☞ Rule:	**Faulty or improper comparisons often leave out key words, particularly demonstrative pronouns such as "those" and "that," which are essential to meaning.**
Incorrect	The attention span of a dolphin is greater than a chimpanzee.
Q65	The attention span of a dolphin is greater than _____ a chimpanzee.
Incorrect	The requirements of a medical degree are more stringent than a law degree.
Q66	The requirements of a medical degree are more stringent than _____ a law degree.
Incorrect	Like many politicians, the senator's promises sounded good but ultimately led to nothing.
Q67	Like _____ many politicians, the senator's promises sounded good but ultimately led to nothing.

☞ Rule: "Like" is used with phrases; "as" is used with clauses. A "phrase" is a group of related words that doesn't have both a subject and a verb; a "clause" is a group of related words that does have a subject and a verb. An easier way to remember the difference is to simply say, "A phrase is a group of words which doesn't have a verb; a clause is a group of words which does have a verb."

Q68   No one hits home runs (as / like) Barry Bonds.

Q69   No one pitches (as / like) Roy Halladay does.

Verb Tenses

EXHIBIT 1.3   THE SIMPLE AND PROGRESSIVE VERB FORMS

	Simple Form	Progressive Form
Present Tense	I travel	I am traveling
Past Tense	I traveled	I was traveling
Future Tense	I will travel	I will be traveling
Present Perfect Tense	I have traveled	I have been traveling
Past Perfect Tense	I had traveled ...	I had been traveling ...
Future Perfect Tense	I will have traveled ...	I will have been traveling ...

Exhibit 1.4 Visualizing the Six Verb Tenses

Tense	Examples	Summary
Simple Present	I study physics.	Expresses events or situations that currently exist, including the near past and near present.
Simple Past	I studied physics.	Expresses events or situations that existed in the past.
Simple Future	I will study physics.	Expresses events or situations that will exist in the future.
Present Perfect	I have studied physics.	Expresses events or situations that existed in the past but that touch the present.
Past Perfect	By the time I graduated from high school, I had decided to study physics.	Expresses events or situations in the past, one of which occurred before the other.
Future Perfect	By the time I graduate from college, I will have studied physics for four years.	Expresses events or situations in the future, one of which will occur after the other.

☞ Rule: Consistent use of verb tenses generally requires that a single sentence be written solely in the present, past, or future tense.

Q70   My dog barks when he (sees / saw) my neighbor's cat.

Q71   Yesterday afternoon, smoke (fills / filled) the sky and sirens sounded.

Q72   Tomorrow, we (will go / will have gone) to the football game.

☞ Rule: The present perfect tense employs the verbs "has" or "have." The past perfect tense employs the auxiliary "had." The future perfect tense employs the verb form "will have."

Q73   We are raising money for the new scholarship fund. So far we (raised / have raised / had raised) $25,000.

Q74   By the time I began playing golf, I (played / had played) tennis for three hours.

Q75   Larry (studied / has studied / had studied) Russian for five years before he went to work in Moscow.

Q76   By the time evening arrives, we (finished / had finished / will have finished) the task at hand.

☞ Rule: The subjunctive mood uses the verb "were" instead of "was." The subjunctive mood is used to indicate a hypothetical situation—it may express a wish, doubt, or possibility. It is also used to indicate a contrary-to-fact situation.

Q77 Sometimes she wishes she (was / were) on a tropical island having a drink at sunset.

Q78 If I (was / were) you, I would be feeling quite optimistic.

☞ Rule: Conditional statements are most commonly expressed in an "If... then" format, in which case an "if" clause is followed by a "results" clause. Confusion often arises as to whether to use "will" or "would." The choice between these verb forms depends on whether a given conditional statement involves the subjunctive. For situations involving the subjunctive, the appropriate verb form is "would"; for situations not involving the subjunctive, the verb form is "will." A helpful hint is that "would" is often used in conjunction with "were"—the appearance of both these words within the same sentence is the telltale sign of the subjunctive.

Q79 If economic conditions further deteriorate, public confidence (will / would) plummet.

Q80 If economic conditions were to further deteriorate, public confidence (will / would) plummet.

Q81 If my taxes are less than $10,000, I (will / would) pay that amount immediately.

Q82 If oil (was / were) still abundant, there (will / would) be no energy crisis.

Diction Review

Diction may be thought of as "word choices." Choose the answer that conforms to the proper use of diction.

Q83 (A) <u>Everyone</u> of the makeup exams is tough, but <u>anyone</u> who misses a scheduled test with good cause is entitled to write one.

(B) <u>Every one</u> of the makeup exams is tough, but <u>anyone</u> who misses a scheduled test with good cause is entitled to write one.

(C) <u>Every one</u> of the makeup exams is tough, but <u>any one</u> who misses a scheduled test with good cause is entitled to write one.

Q84 (A) The green book, <u>that</u> is on the top shelf, is the one you need for math. The book <u>which</u> is red is the one you need for writing.

(B) The green book, <u>which</u> is on the top shelf, is the one you need for math. The book <u>that</u> is red is the one you need for writing.

(C) The green book, <u>which</u> is on the top shelf, is the one you need for math. The book <u>which</u> is red is the one you need for writing.

Q85 (A) <u>Let's</u> cherish the poem "In Flanders Fields." Remembering those who fought for our freedom <u>lets</u> us live easier.

(B) <u>Lets</u> cherish the poem "In Flanders Fields." Remembering those who fought for our freedom <u>let's</u> us live easier.

(C) <u>Let's</u> cherish the poem "In Flanders Fields." Remembering those who fought for our freedom <u>let's</u> us live easier.

Q86 (A) Once we turn these dreaded assignments <u>into</u> the professor's office, we'll feel a lot less obliged to pass any information <u>onto</u> our classmates.

(B) Once we turn these dreaded assignments <u>into</u> the professor's office, we'll feel a lot less obliged to pass any information <u>on to</u> our classmates.

(C) Once we turn these dreaded assignments <u>in to</u> the professor's office, we'll feel a lot less obliged to pass any information <u>on to</u> our classmates.

Q87 (A) The McCorkendales didn't <u>used to</u> enjoy warm weather, but that was before they moved to Morocco and got <u>used to</u> summer temperatures as high as 35 degrees Celsius.

(B) The McCorkendales didn't <u>use to</u> enjoy warm weather, but that was before they moved to Morocco and got <u>use to</u> summer temperatures as high as 35 degrees Celsius.

(C) The McCorkendales didn't <u>use to</u> enjoy warm weather, but that was before they moved to Morocco and got <u>used to</u> summer temperatures as high as 35 degrees Celsius.

Idioms Review

Idioms may be thought of as "word expressions." Idioms, like grammar and diction, are correct or incorrect, right or wrong. Here are fifteen common idioms.

Q88 (A) A choice must be made <u>between</u> blue <u>and</u> green.

(B) A choice must be made <u>between</u> blue <u>or</u> green.

Q89 (A) Many doctors <u>consider</u> stress a more destructive influence on one's longevity than smoking, drinking, or overeating.

(B) Many doctors <u>consider</u> stress <u>as</u> a more destructive influence on one's longevity than smoking, drinking, or overeating.

(C) Many doctors <u>consider</u> stress <u>to be</u> a more destructive influence on one's longevity than smoking, drinking, or overeating.

Q90 (A) At first women were <u>considered</u> at low risk for HIV.

(B) At first women were <u>considered as</u> at low risk for HIV.

(C) At first women were <u>considered to be</u> at low risk for HIV.

Q91 (A) Many <u>credit</u> Gutenberg <u>as having</u> invented the printing press.

(B) Many <u>credit</u> Gutenberg <u>with having</u> invented the printing press.

Q92 (A) In the movie *Silence of the Lambs*, Dr. Hannibal Lecter is <u>depicted as</u> a brilliant psychiatrist and cannibalistic serial killer who is confined as much by the steel bars of his cell as by the prison of his own manufacture.

(B) In the movie *Silence of the Lambs*, Dr. Hannibal Lecter is <u>depicted to be</u> a brilliant psychiatrist and cannibalistic serial killer who is confined as much by the steel bars of his cell as by the prison of his own manufacture.

Q93 (A) Only experts can <u>distinguish</u> a masterpiece <u>and</u> a fake.

(B) Only experts can <u>distinguish</u> a masterpiece <u>from</u> a fake.

Q94 (A) Although medical practitioners have the technology to perform brain transplants, there is no clear evidence that they can <u>do it</u>.

(B) Although medical practitioners have the technology to perform brain transplants, there is no clear evidence that they can <u>do so</u>.

Q95 (A) <u>In comparison to</u> France, Luxembourg is an amazingly small country.

(B) <u>In comparison with</u> France, Luxembourg is an amazingly small country.

Q96 (A) Pete Sampras won Wimbledon with a classic tennis style, <u>in contrast to</u> Bjorn Borg, who captured his titles using an unorthodox playing style.

(B) Pete Sampras won Wimbledon with a classic tennis style, <u>in contrast with</u> Bjorn Borg, who captured his titles using an unorthodox playing style.

Q97 (A) There is <u>more</u> talk of a single North American currency today <u>compared to</u> ten years ago.

(B) There is <u>more</u> talk of a single North American currency today <u>compared with</u> ten years ago.

(C) There is <u>more</u> talk of a single North American currency today <u>than</u> ten years ago.

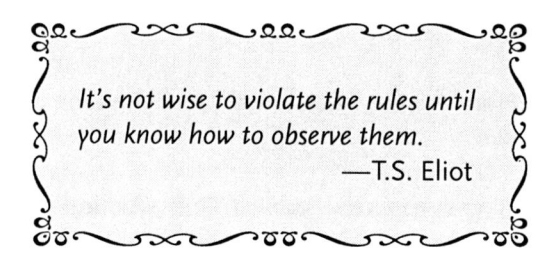

THE 100-QUESTION QUIZ

Q98   (A)   I <u>prefer</u> blackjack <u>over</u> poker.

        (B)   I <u>prefer</u> blackjack <u>to</u> poker.

Q99   (A)   Rembrandt is <u>regarded as</u> the greatest painter of the Renaissance period.

        (B)   Rembrandt is <u>regarded to be</u> the greatest painter of the Renaissance period.

Q100   (A)   The author does a good job of <u>tying</u> motivational theory <u>to</u> obtainable results.

        (B)   The author does a good job of <u>tying</u> motivational theory <u>with</u> obtainable results.

> *It's not wise to violate the rules until you know how to observe them.*
> —T.S. Eliot

Answers to the 100-Question Quiz

Q1   An office clerk and a machinist <u>were</u> present but unhurt by the on-site explosion.

Q2   Eggs and bacon <u>is</u> Tiffany's favorite breakfast.

The words "eggs" and "bacon" are intimately connected and deemed to be a signal unit.

Q3   A seventeenth-century oil painting, along with several antique vases, <u>has</u> been placed on the auction block.

Q4   The purpose of the executive, administrative, and legislative branches of government <u>is</u> to provide a system of checks and balances.

(The subject of the sentence is "purpose." The prepositional phrase "of the executive, administrative, and legislative branches of government" does not affect the verb choice.)

Q5   Here <u>are</u> the introduction and chapters one through five.

(The compound subject "introduction *and* chapters one through five" necessitates using the plural verb "are.")

Q6   <u>Are</u> there any squash courts available?

One helpful tip is to first express this as a declarative sentence: "There are squash courts available." Now it is easier to see that the subject is plural—squash courts—and a plural verb *are* is appropriate.

Q7   Entertaining multiple goals <u>makes</u> a person's life stressful.

"Entertaining multiple goals" is a gerund phrase which acts as the subject of the sentence (singular).

Q8 To plan road trips to three different cities <u>involves</u> the handling of many details.

"To plan roads trips" is an infinitive phrase which acts as the subject of the sentence (singular).

Q9 One in every three new businesses <u>fails</u> within the first five years of operation.

Q10 Few of the students, if any, <u>are</u> ready for the test.

The phrase "if any" is parenthetical, and in no way affects the plurality of the sentence.

Q11 Some of the story <u>makes</u> sense.

Q12 Some of the comedians <u>were</u> hilarious.

Q13 None of the candidates <u>have</u> any previous political experience.

Note that if "neither" was used in place of "none," the correct sentence would read: "Neither of the candidates <u>has</u> any political experience." "Neither" is an indefinite pronoun that is always singular. "None" is an indefinite pronoun that is singular or plural depending on context. The fact that "none" takes "have" and "neither" would take "has" is indeed a peculiarity.

Q14 Either Johann or Cecilia <u>is</u> qualified to act as manager.

Q15 Neither management nor workers <u>are</u> satisfied with the new contract.

Q16   Our group is meeting at 6 p.m.

Q17   A group of latecomers were escorted to their seats.

Q18   The number of road accidents has decreased.

Q19   A number of train accidents have occurred.

Q20   Fifty percent of video gaming is having great reflexes.

Q21   Two-thirds of their classmates have wakeboards.

Q22   Ten dollars is an average daily wage for many people in the developing word.

Q23   The present is from Beth and her.

Q24   Cousin Vinny and he are both valedictorians.

Q25   Between you and me, this plan makes a lot of sense.

The pronoun "me" (the objective form of the pronoun "I") is the direct object of the preposition "between."

Q26   Do not ask for whom the bell tolls.

The pronoun "whom" (the objective form of the pronoun "who") is the direct object of the preposition "for."

Q27   People like you and me should know better.

The objective form of the pronoun—"me"—must follow the preposition "like."

Q28   My son is more sports minded than I.
In order to test this: My son is more sports minded than I am.

Q29   We skate as fast as they.

Test this: We skate as fast as they do.

Q30   During our group presentation, our teacher asked you more questions than me.

Test this: During our group presentation, our teacher asked you more questions than she or he asked me.

Q31   The woman who is responsible for pension planning is Mrs. Green.

She is responsible for city planning; "he/she" is substitutable for "who."

Q32   This gift is intended for whom?

The gift is intended for *him* or *her*; "him/her" is substitutable for "whom."

Q33   The tour leader told Julie and me to turn our cell phones off.

Q34   Young Robert hurt himself while climbing alone.

Q35   A not-for-profit, like any other organization, has its own rules and regulations to follow.

Q36   Everybody should mind his or her own business.

Q37   Sam never argues with his father when Sam is drunk.

The sentence "Sam never argues with his father when he is drunk" is grammatically correct but contextually vague. It is contextually vague because we feel that it is Sam who is drunk whereas, grammatically, it is Sam's father who is actually drunk (a pronoun modifies the nearest noun that came before it; here the pronoun "he" modifies the noun "father"). The sentence needs to be rephrased to clear up potential ambiguity. The most direct way to achieve this is to replace the pronoun "he" with the noun it is intended to refer to, namely Sam. Note that another way to clear up this ambiguity is to restructure this sentence as follows: "When he is drunk, Sam never argues with his father."

Q38   To know that a person can't vote is to know that <u>he or she doesn't</u> have a voice.

A "person" is a noun in the third person and the correct answer must be a pronoun that matches it in the third person.

Other correct options would include:

To know that a person can't vote is to know that <u>a person doesn't</u> have a voice.

To know that a person can't vote is to know that <u>one doesn't</u> have a voice.

Q39   One cannot really understand another country until <u>one has</u> studied its history and culture.

We have essentially five ways to validate this sentence—"one has," "a person has," "he has," "she has," or "he or she has." In the latter option, using "he or she has" keeps writing gender neutral (politically correct). The grammatical reason that the original does the work is because "one" is a third-person singular pronoun while "they" is a third-person plural pronoun. Thus, we have a pronoun shift or a shift in viewpoint. Any answer must also be in the third-person singular. Given the opportunity to rewrite the original sentence, two other correct options would also include:

You cannot really understand another country unless <u>you have</u> studied its history and culture.

Here, the second-person pronoun "you" is matched with the second-person pronoun "you."

We cannot really understand another country unless <u>we have</u> studied its history and culture. Here the first-person plural pronoun "we" is matched with the first-person plural pronoun "we."

Q40 He <u>frequently</u> told her he wanted to marry her.

Q41 The janitor was surprised to find termites <u>coming out of the wood</u>.

Q42 After writing the introduction, <u>I</u> easily drafted the rest of the report.

Q43 Walking along the shore, <u>the couple</u> could see fish jumping in the lake.

Q44 She said she had <u>a copy of the map</u> lying in her office.

She is presently not in her office but the map is.

Also: While we were sitting in her office, she told me she had a copy of the map.

She is in her office with or without the map.

Q45 In addition to building organizational skills, <u>I also honed my team-building skills during the summer internship</u>.

Q46   An incredibly complex mechanism, <u>the brain</u> has some 10 billion nerve cells.

Q47   <u>On the basis of</u> our observations, we believe the project will succeed.

Firstly, "the project" is not based on our observations. Observations must be made by people, so "we" is an appropriate substitute. Secondly, the phrase "based on" is incorrect because we cannot be physically standing on our observations or attached to them. The correct phraseology is "on the basis of." In general, "based on" is not an appropriate modifier to use with people; but it's fine for inanimate objects, e.g., a movie based on a book.

Q48   In the summer before college, Max <u>waited</u> tables, <u>sold</u> magazines, and even <u>delivered</u> pizzas.

Q49   Our neighbors went <u>to</u> London, Athens, and Rome.

Q50   Our neighbors went <u>to</u> London, <u>to</u> Athens, and <u>to</u> Rome.

Q51   Jonathan likes <u>not only</u> rugby but also kayaking.

Here the verb "likes" is placed before the *not only ... but also* correlative conjunction, creating parallelism between the words "rugby" and "kayaking."

Q52   Jonathan <u>not only likes</u> rugby <u>but also likes</u> kayaking.

Here, the verb "likes" is repeated after each component part of the *not only ... but also* construction. Thus the words "likes rugby" and "likes kayaking" are parallel.

Q53 To examine the works of William Shakespeare—his plays and poetry—is to marvel at one man's seemingly incomparable depth of literary expression.

The infinitives "to examine" and "to marvel" are parallel.

Q54 In the *Phantom of the Opera* play, the story is intriguing and the singers are superb.

Since the verbs are different (i.e., "is" and "are"), we must write them out.

NOTE ෬ Rules of ellipsis govern the acceptable omission of words in writing and speech. There is no need to say, "Paris is a large and is an exciting city." The verb (i.e., "is") is the same throughout the sentence, so there's no need to write it out. Note, however, that the articles "a" and "an" are different and must be written out. Omitting the "an" in the second half of the sentence would result in the nonsensical, "Paris is exciting city."

Q55 The defendant's own testimony on the stand neither contributed to nor detracted from his claim of innocence.

Since the prepositions are different, we cannot omit either of them

NOTE ෬ As a follow-up example, there is no need to say, "*The Elements of Style* was written by William Strunk, Jr., and was written by E. B. White." Since the verb form "was written" and the preposition "by" are the same when applied to both authors, we can simply say, "*The Elements of Style* was written by William Strunk, Jr., and E. B. White."

Q56   Between Tom and Brenda, Tom is <u>better</u> at math.

Q57   Among our group, Jeff is the <u>wealthiest</u> person.

Q58   Of all the roses grown in our neighborhood, Chauncey Gardiner's grow the <u>most</u> vigorously.

Q59   Chauncey Gardiner's roses grow <u>more</u> vigorously than any other in the neighborhood.

Q60   Tokyo's population is greater than the <u>population</u> of Beijing.

Q61   Tokyo's population is greater than Beijing's <u>population</u>.

Q62   Tokyo's population is greater than that of <u>Beijing</u>.

In the above example, the demonstrative pronoun "that" substitutes for the words "the population," and we are effectively saying: "Tokyo's population is greater than <u>the population</u> of Beijing."

NOTE ☙ It is incorrect to write: "Tokyo's population is greater than <u>that of Beijing's</u>." Such a sentence would read: "Tokyo's population is greater than the population of Beijing's (population)."

Q63   Tokyo's population is greater than <u>Beijing's</u>.

Also: Tokyo's population is greater than <u>Beijing's population</u>.

Also: Tokyo's population is greater than <u>that of Beijing</u>.

Also: Tokyo's population is greater than <u>the population of Beijing</u>.

Q64 Of all the countries contiguous to India, <u>Pakistan has the most strongly defended borders</u>.

The following would <u>not</u> be a correct solution: "Of all the countries contiguous to India, <u>the borders of Pakistan are most strongly defended</u>."

Q65 The attention span of a dolphin is greater than <u>that of</u> a chimpanzee.

Q66 The requirements of a medical degree are more stringent than <u>those of</u> a law degree.

Q67 Like <u>those of</u> many politicians, the senator's promises sounded good but ultimately led to nothing.

Alternatively, we could use the words "like the promises of" in the following manner: "Like <u>the promises of</u> many politicians, the senator's promises sounded good but ultimately led to nothing." Ignoring the fill-in-the-blank, we could also write: "Like many politicians' promises, the senator's promises..."

Q68 No one hits home runs <u>like</u> Barry Bonds.

"Like Barry Bonds" is a phrase. A phrase is a group of words that lacks a verb.

Q69 No one pitches <u>as</u> Roy Halladay does.

"As Roy Halladay does" is a clause. A clause is a group of words that contains a verb.

Q70 My dog barks when he <u>sees</u> my neighbor's cat.

The simple present tense "barks" is consistent with the simple present tense "sees."

Q71   Yesterday afternoon, smoke <u>filled</u> the sky and sirens sounded.

The simple past tense verb "filled" is consistent with the simple past tense verb "sounded."

Q72   Tomorrow, we <u>will go</u> to the football game.

Q73   We are raising money for the new scholarship fund. So far we <u>have raised</u> $25,000.

Q74   By the time I began playing golf, I <u>had played</u> tennis for three hours.

The playing of tennis precedes the playing of golf for these two past tense events.

Q75   Larry <u>had studied</u> Russian for five years before he went to work in Moscow.

There are actually two correct answers here. Sentence 1 correctly employs the past perfect tense.

1)   Larry <u>had studied</u> Russian for five years before he went to work in Moscow.

The past perfect tense is constructed using the auxiliary "had" and the past participle of the verb, in this case "studied." The past perfect tense clarifies (or helps to clarify) the sequence of two past tense events. Here, it is clear that Larry first studied Russian and then went to Moscow.

Sentence 2 correctly uses two past tense verbs (i.e., "studied" and "went") as well as the temporal word "before."

2) Larry studied Russian for five years before he went to work in Moscow.

Temporal words are words indicating time—their more technical name is "temporal conjunctions." When the sequence of two past tense events is clear, particularly through the use of temporal words (e.g., *before, after, previously, prior, subsequently*), the use of the past perfect tense is considered optional. Some experts claim that the combined use of the past perfect "had" and temporal words, as seen in sentence 1 above, creates a redundancy. Other experts side with sentence 1 as the preferred answer, even though confirming that both scenarios are correct. Most grammar books side with sentence 1 as the correct and preferable choice.

NOTE ☙ To clear up some of the confusion surrounding use of the word "had," let's review its main uses. First, "had" is used as an actual verb and functions as the past tense of the verb "to have." Examples: "I have 500 dollars" versus "I had 500 dollars." In the previous example, "had" is a verb meaning "to possess." It also functions as a verb meaning "to experience" or "to undergo." Examples: "I had a good time at the party" or "I had a bad headache." It also functions as a verb meaning "to be required to." Example: "I had to go to the store today to get some medicine for my mother."

As already mentioned, one important use of the auxiliary "had" is to form the past perfect tense. Additionally, the auxiliary "had" can also play a role in forming the subjunctive ("I wish I had done things differently") and the conditional ("If I had known then what I know now, things might have been different.")

Confusion may arise in situations involving the use of "had had." The past perfect is formed by using the auxiliary "had" plus the past participle of a verb. In situations involving the verb "had," the past perfect tense becomes "had had." Example: "By the time he turned twenty-five, he had had six different jobs." Here, the act of working at six different jobs occurs prior to turning twenty-five years of age, and the past perfect tense is invoked to clarify the sequence of events. One way to avoid employing two "had's" is to change the verb, when applicable. In this case, we could write, "By the time he turned twenty-five, he had worked at six different jobs." The verb (past participle) "worked" substitutes for the verb "had."

One common mistake is to place "had" before past tense verbs. For example:

1) I have lived in Toronto.
2) I lived in Toronto.
3) I had lived in Toronto.

Statement 3—"I had lived in Toronto"—is not grammatically correct (at least not standing on its own). "Had lived" is not the past of "have lived." The past tense of "have lived" is "lived." In other words, the past of the present perfect tense is the simple past, not the past perfect tense. Only statements 2 and 3 above are grammatically correct. In summary, although the verb "had" is, in fact, the past tense of the verb "(to) have," "had lived" is not the past tense of "have lived." The truth is that many everyday writers now associate "had" with the past tense, and thus "sprinkle" it in front of many past tense verbs.

Let's examine further the ubiquitous practice of placing "had" in front of past tense verbs, however objectionable. Consider the following two statements:

1) "He <u>worked</u> in the diplomatic corps."
2) "He <u>had worked</u> in the diplomatic corps."

Only the first statement is grammatically correct. The second statement (standing alone), although colloquial, is not grammatically sound. Statement 1 is the simple past tense. He worked in the diplomatic corps for a specific period of time in the past, but doesn't anymore. Statement 2, to the casual ear, carries a meaning nearly identical to statement 1. It appears that many everyday writers prefer statement 2 to statement 1, a likely reason being that it may sound better to one's ears.

A good rule of thumb is to omit the use of "had" if it isn't needed. Instead of writing "I <u>had</u> thought a lot about what you said," write "I thought a lot about what you said." A past tense verb does not need the help of the auxiliary "had" to do its job. That said, since so many writers now associate "had" with the simple past tense, the practice of placing "had's" in front of past tense verbs is likely so entrenched that the practice is here to stay.

Another common practice is to "invoke" the past perfect tense even though the simple past tense is called for. Consider the following pairings:

1) They <u>went</u> to Santa Catalina Island many times.
2) They <u>had gone</u> to Santa Catalina Island many times.

1) She <u>grew</u> her hair long.
2) She <u>had grown</u> her hair long.

1) He <u>was</u> a civil servant.
2) He <u>had been</u> a civil servant.

All three of the previous word pairings sound very much equivalent. It is essentially a draw between the simple past tense ("went" or "grew" or "was") and the auxiliary "had" + past participle ("had gone" or "had grown" or "had been"). The point here is that it is understandable why writers might choose to invoke the past perfect tense even though there are no grammatical grounds for doing so. As a practical matter, writers should feel free to use whatever form sounds better. To be clear, each of these "second" sentences could represent legitimate examples of the past perfect tense given additional context. Cases in point: "Before moving to Oregon, they had gone to Santa Catalina Island many times"... "By the time she entered high school, she had grown her hair long"... "He had been a civil servant until deciding to start his own business."

A close cousin of the "invoked" past perfect may be called the "invoked" present perfect. Consider the following trio:

1) I have misplaced my car keys.
2) I misplaced my car keys.
3) I had misplaced my car keys.

Statement 2 is the simple past tense and it could be argued, on grounds of logic, that it represents the only grammatically sound statement. Statement 3 is an example of the "invoked" past perfect tense; it is in common use but, as mentioned, is arguably not technically correct. Statement 1 is the present perfect tense. The question becomes: What does statement 1 really mean? In reality, either I misplaced my car keys or I didn't misplace my car keys. If I really did misplace my car keys, then why isn't statement 1, written in the simple past tense, sufficient to express this idea?

Many writers likely "invoke" the present perfect tense, as seen in statement 2, because it sounds right. This may be an apt illustration of how the written English language is being influenced by the way we speak it and hear it being spoken. The use of "have misplaced" (present perfect tense) likely makes the event seem as if it occurred only a short time ago. The use of "misplaced" (past tense) makes the event seem as if occurred at a time further in the past. The use of "had misplaced" (past perfect tense) makes it seem as if the event occurred at a time in the yet more distant past.

In summary, as long as we're referring to an indefinite point or time period in the past, all of these three variations, as found in everyday writing and speech, are largely interchangeable.

Q76 By the time evening arrives, we <u>will have finished</u> the task at hand.

The future act of finishing the task at hand will occur before evening arrives.

Q77 Sometimes she wishes she <u>were</u> on a tropical island having a drink at sunset.

Expresses a wish; the subjunctive "were," not "was," is the correct choice.

Q78 If I <u>were</u> you, I would be feeling quite optimistic.

Indicates a hypothetical, contrary-to-fact situation; "were," not "was," is the correct choice.

Q79 If economic conditions further deteriorate, public confidence <u>will</u> plummet.

"Will" is correct in future events with implied certainty; we are making a statement about the future in absolute terms. The sentence is written in the form of "If *x* happens, then *y* will happen."

Q80   If economic conditions were to further deteriorate, public confidence would plummet.

Note that the inclusion of "were," when coupled with "would," signals the subjunctive mood.

Q81   If my taxes are less than $10,000, I will pay that amount immediately.

"Will" is correct when dealing with future events with implied certainty.

Q82   If oil were still abundant, there would be no energy crisis.

This situation is clearly contrary to fact. Oil is not abundant, and there is an energy crisis; "were" and "would" are used to signal the subjunctive.

Q83   Choice B
Every one of the makeup exams is tough, but anyone who misses a scheduled test with good cause is entitled to write one.

The words *anyone* and *any one* are not interchangeable. *Anyone* means "any person" whereas *any one* means "any single person or thing." Likewise, the words *everyone* and *every one* are not interchangeable. *Everyone* means "everybody in a group" whereas *every one* means "each person."

Q84 Choice B
The green book, <u>which</u> is on the top shelf, is the one you need for math. The book <u>that</u> is red is the one you need for grammar.

It is common practice to use *which* with non-restrictive (nonessential) phrases or clauses and to use *that* with restrictive (essential) phrases or clauses. Nonrestrictive phrases are typically enclosed with commas, whereas restrictive phrases are never enclosed with commas. "Which is on the top shelf" is a nonrestrictive (nonessential) phrase. It is optional. We can omit it, and the sentence will still make sense. "That is red" is a restrictive (essential) phrase. It is not optional. Without it the sentence will not make sense.

Q85 Choice A
<u>Let's</u> cherish the poem "In Flanders Fields." Remembering those who fought for our freedom <u>lets</u> us live easier.

Let's is a contraction for "let us"; *lets* is a verb meaning "to allow" or "to permit." This sentence could have been rewritten: <u>Let us</u> cherish the poem "In Flanders Fields." Remembering those who fought for our freedom *allows* us to live easier.

Q86 Choice C
Once we turn these dreaded assignments <u>in to</u> the professor's office, we'll feel a lot less obliged to pass information <u>on to</u> our classmates.

The words *into* and *in to* are not interchangeable. Likewise, the words *onto* and *on to* are not interchangeable. Case in point: Turning assignments *into* the professor's office is a magician's trick! Passing information *onto* our classmates would mean physically putting the information on them.

Q87   Choice C
The McCorkendales didn't <u>use to</u> fancy warm weather, but that was before they moved to Morocco and got <u>used to</u> summer temperatures as high as 35 degrees Celsius.

Although *used to* and *use to* are largely interchangeable in spoken English, because the letter "d" is inaudible in many oral contexts, this is not the case in formal writing. The correct form for habitual action is *used to*, not *use to*. Example: "We <u>used to</u> go to the movies all the time." However, when *did* precedes "use(d) to" the correct form is "use to." This is commonly the case in questions and negative constructions. Example: Didn't you <u>use to</u> live on a farm? I didn't <u>use to</u> daydream.

Q88   Choice A
Idiom: *Between X and Y*

A choice must be made <u>between</u> blue <u>and</u> green.

Q89   Choice A
Idiom: *Consider(ed)* – not followed by *"to be"*

Many doctors <u>consider</u> stress a more destructive influence on one's longevity than smoking, drinking, or overeating.

Consider/considered is not followed by "to be" (or "as") when *consider(ed)* is followed by a direct object and used in the sense that some person or organization considers something to have some perceived quality. The word "stress" functions as a direct object of the verb *consider*, and the perceived quality is the "destructive influence" of stress.

Q90 Choice C
Idiom: *Consider(ed)* – followed by *"to be"*

At first women were considered to be at low risk for HIV.

Consider/considered is followed by "to be" when *consider(ed)* has the meaning of "believed to be" or "thought to be."

Q91 Choice B
Idioms: *Credit(ed) X with having*

Many credit Gutenberg with having invented the printing press.

Q92 Choice A
Idiom: *Depicted as*

In the movie *Silence of the Lambs,* Dr. Hannibal Lecter is depicted as a brilliant psychiatrist and cannibalistic serial killer who is confined as much by the steel bars of his cell as by the prison of his own manufacture.

Q93 Choice B
Idiom: *Distinguish X from Y*

Only experts can distinguish a masterpiece from a fake.

Q94 Choice B
Idiom: *Do so*

Although doctors have the technology to perform brain transplants, there is no clear evidence that they can do so.

Q95   Choice A
Idiom: *In comparison to*

<blockquote>In comparison to France, Luxembourg is an amazingly small country.</blockquote>

Q96   Choice A
Idiom: *In contrast to*

<blockquote>Pete Sampras won Wimbledon with a classic tennis style, in contrast to Bjorn Borg, who captured his titles using an unorthodox playing style.</blockquote>

Q97   Choice C
Idiom: *More…than/(Less…than)*

<blockquote>There is more talk of a single North American currency today than ten years ago.</blockquote>

Q98   Choice B
Idiom: *Prefer X to Y*

<blockquote>I prefer blackjack to poker.</blockquote>

Q99   Choice A
Idiom: *Regarded as*

<blockquote>Rembrandt is regarded as the greatest painter of the Renaissance period.</blockquote>

Q100   Choice A
Idiom: *Tying X to Y*

<blockquote>The author does a good job of tying motivational theory to obtainable results.</blockquote>

Chapter 2
Grammatical Munchkins

This chapter contains a repository of grammatical terms that serves as a glossary, replete with clarifying examples. The value of this section lies in its ability to provide a convenient summary of important terms as well as a jumping-off point—a means of engendering interest—for those readers wishing to pursue more in-depth study.

The Eight Parts of Speech

There are eight parts of speech in English: nouns, pronouns, verbs, adjectives, adverbs, prepositions, conjunctions, and interjections.

Noun A noun is a word that names a person, place, thing, or idea.

Example: <u>Sally</u> is a nice person and you can speak freely with her.

Pronoun A pronoun is a word used in place of a noun or another pronoun.

Example: Sally is a nice person and <u>you</u> can speak freely with <u>her</u>.

Verb A verb is a word that expresses an action or a state of being.

Example: Sally <u>is</u> a nice person and you <u>can speak</u> freely with her.

Adjective An adjective is a word used to modify or describe a noun or pronoun.

Example: Sally is a <u>nice</u> person and you can speak freely with her. The adjective "nice" modifies the noun "person."

Adverb An adverb is a word that modifies an adjective, a verb, or another adverb.

Example: Sally is a nice person and you can speak <u>freely</u> with her. The word "freely" modifies the verb "speak."

Preposition

A preposition is a word that shows a relationship between two or more words.

Example: Sally is a nice person and you can speak freely <u>with</u> her.

Prepositions are sometimes informally referred to as words that describe "the directions a squirrel can go." Squirrels, after all, seem to be able to run, climb, or crawl in nearly every possible direction.

Examples of prepositions include: *after, against, at, before, between, by, concerning, despite, down, for, from, in, of, off, on, onto, out, over, through, to, under, until, up, with.*

Conjunction

A conjunction is a word that joins or connects words, phrases, clauses, or sentences. Three major types of conjunctions include coordinating conjunctions, subordinating conjunctions, and correlative conjunctions.

Example: Sally is a nice person <u>and</u> you can speak freely with her.

Interjection

An interjection is a word or a term that denotes a strong or sudden feeling. Interjections are usually, but not always, followed by an exclamation mark.

Example: Sally is a nice person and you can speak freely with her. <u>Wow!</u>

Parts of Speech vs. The Seven Characteristics

Each of the eight parts of speech have one or more of the following characteristics: (1) gender, (2) number, (3) person, (4) case, (5) voice, (6) mood, and (7) tense. The matching of a particular part of speech with its relevant characteristics is the primary "cause" of grammar.

NOTE ଔ Adjectives, adverbs, prepositions, conjunctions, and interjections do not have gender, number, person, case, voice, mood, or tense. Only nouns, pronouns, and verbs have one or more of these seven characteristics.

Gender   Gender may be feminine or masculine. Only nouns and pronouns have gender.

Examples: Masculine—*boy* (noun), *him* (pronoun). Feminine—*girl* (noun), *her* (pronoun).

Number   Number may be singular or plural. Only nouns, pronouns, and verbs have number.

Examples: Singular—*home* (noun), *I* (pronoun), *plays* (verb). Plural—*homes* (noun), *we* (pronoun), *play* (verb).

Person   Person may be first person, second person, or third person. A person doing the speaking is considered first person; the person spoken to is considered second person; a person spoken about is considered third person. Only pronouns and verbs have person.

Examples: First person—*I write* (pronoun + verb). Second person—*you write* (pronoun + verb). Third person—*he writes* (pronoun + verb).

NOTE ☙ When verbs are matched with personal pronouns, verbs differ only in number with respect to third-person singular pronouns. In the third-person singular, verbs are formed with the letter "s." For example: "He or she travels." But: "I travel," "you travel," and "they travel."

Case

Case may be subjective, objective, or possessive. Only nouns and pronouns have case.

Examples: Subjective—*Felix has a cat* ("Felix" is a noun); *He has a cat* ("he" is a pronoun). Objective—*The cat scratched Felix* ("Felix" is a noun); *The cat scratched him* ("him" is a pronoun). Possessive—*Felix's cat has amber eyes* (Felix's is a noun); *His cat has amber eyes* ("his" is a pronoun).

NOTE ☙ Although nouns have case, noun forms remain virtually unchanged in the subjective, objective, and possessive cases.

Voice

Voice may be active or passive. Only verbs have voice.

Examples: Active voice—*You mailed a letter.* Passive voice—*The letter was mailed by you.*

In the active voice, the doer of the action is placed at the front of the sentence;

the receiver of the action is placed at the back of the sentence. In the passive voice, the receiver of the action is placed at the front of the sentence while the doer of the action is relegated to the back of the sentence.

Mood

Mood can be described as being indicative, imperative, or subjunctive. Only verbs have mood.

Examples: Indicative mood (makes a statement or asks a question)—*It's a nice day*. Imperative mood (makes a request or gives a command)—*Please sit down*. Subjunctive mood (expresses a wish or a contrary-to-fact situation)—*I wish I were in Hawaii*.

Tense

Tense refers to time. There are six tenses in English—present tense, past tense, future tense, present perfect tense, past perfect tense, and future perfect tense. Each of these six tenses occurs within two forms: the simple form and the progressive form.

Examples: Present tense in the simple form—*I study*. Present tense in the progressive form—*I am studying*.

Other Grammatical Terms

Adjective clause	An adjective clause is a subordinate clause that, like an adjective, modifies a noun or pronoun.
	Example: "The house that sits on top of the hill is painted in gold." The adjective clause "that sits on top of the hill" describes the "house."
Antecedent	An antecedent is the word to which a pronoun refers. It is the word that the pronoun is effectively taking the place of.
	Example: "The clock is broken; it is now being repaired." The pronoun "it" is substituting for the antecedent "clock."
Appositive phrase	An appositive phrase is used merely for description and is typically set off by commas.
	Example: The world's oldest book, which was discovered in a tomb, is 2,500 years old.
Article	An article serves to identify certain nouns. English has three articles: *a, an, the. The* is known as a definite article; *a* and *an* are known as indefinite articles. Articles are often erroneously referred to as one of the eight parts of speech.
Clause	A clause is a group of related words that does have a subject and a verb.

Example: "Many people believe in psychics even though they never hear of a psychic winning the lottery." The previous sentence contains two clauses. The first clause—"many people believe in psychics"—is an independent clause, containing the subject "people" and the verb "believe." The second clause—"even though they never hear of a psychic winning the lottery"—is a dependent clause, containing the subject "they" and the verb "winning."

Collective noun

Collective nouns are nouns which represent a group.

Examples: *audience, band, bunch, class, committee, couple, crowd, family, group, herd, jury, majority, people, percent, personnel, team.*

Complement

A complement is something that completes a subject and verb. Not all sentences have complements.

Examples: *I am*—This three-letter sentence (incidentally the shortest in the English language) does not contain a complement. *I am fit*—This sentence does contain a complement; the complement is the word "fit."

Coordinating conjunction

Coordinating conjunctions join clauses of equal weight.

Examples: There are seven coordinating conjunctions in English—*and, but, yet, or, nor, for,* and *so.*

Correlative conjunction	Correlative conjunctions join clauses or phrases of equal weight. They also impose a sense of logic. Examples: *either…or, neither…nor, not only…but (also), both…and,* and *whether…or.* The word appearing in brackets above is deemed optional.
Demonstrative pronoun	Demonstrative pronouns serve to point out persons or things. Example: There are four demonstrative pronouns in English: *this, that, these,* and *those.*
Dependent clause	An dependent clause is a clause that cannot stand on its own as a complete sentence. Dependent clauses are sometimes called subordinate clauses. Example: "Keep an umbrella with you because it's forecast for rain." The dependent clause is "because it's forecast for rain."
Direct object	A direct object (of a verb) receives the action of that verb or shows the result of that action. Example: "The outfielder caught the ball." The word "ball" is the direct object of the verb "caught." See also Indirect Object.
Gerund	Gerunds are verb forms that end in "ing" and function as nouns. Informally they

may be referred to as "words that look like verbs but function as nouns."

Examples: Eating vegetables is good for you. Learning languages is rewarding. Seeing is believing. "Eating," "learning," "seeing," and "believing" are all gerunds.

Indefinite pronoun   Indefinite pronouns are pronouns that do not refer to a specific antecedent.

A more complete list of indefinite pronouns includes: *all, any, anybody, anyone, anything, both, each, either, every, everybody, everyone, everything, few, many, most, neither, nobody, none, no one, nothing, one, several, some, somebody, someone,* and *something.*

Independent clause   An independent clause is a clause that can stand on its own as a complete sentence. Independent clauses are sometimes called main clauses.

Example: "I'm going to carry an umbrella with me because the forecast is for rain." The independent clause is "I'm going to carry an umbrella with me" while the subordinate clause is "because the forecast is for rain."

Indirect object   An indirect object (of a verb) precedes the direct object and usually tells to whom or for whom the action of that verb is done.

Example: "The maître d' gave us a complimentary bottle of wine." The word "us" functions as the indirect object, even though it comes before the direct object. The words "bottle of wine" serve as the direct object.

See also Direct Object.

Infinitive

Infinitives are verb forms, in which the basic form of a verb is preceded by "to." Infinitives generally function as nouns but may also function as adjectives or adverbs. Informally they may be referred to as word pairings in which the preposition "to" is placed in front of a verb.

Examples: To see is to believe. ("To see" and "to believe" are both infinitives.)

Interrogative pronoun

Interrogative pronouns are used in questions.

Examples: *who, which, what, whom,* and *whose.*

Intransitive verb

Intransitive verbs do not require an object to complete their meaning.

Example. He waits. The verb "waits" does not require an object to complete its meaning.

See also Transitive Verb.

Nonrestrictive clause

A nonrestrictive clause is a clause that is not essential to the meaning of a sentence. Nonrestrictive clauses are generally enclosed by commas.

Example: The green book, which is on the top shelf, is the one you need for math class. "Which is on the top shelf" is a nonrestrictive clause.

NOTE ☙ In choosing between "that" or "which," it is common practice to use "that" with restrictive (essential) phrases and clauses and "which" with nonrestrictive (nonessential) phrases and clauses. For this reason, "that" is used with clauses that are not set off by commas and to use "which" is used with clauses that are set off by commas.

See also Restrictive Clauses.

Object

An object (of a verb) is a word or words that receives the action of a verb. An object is a special kind of complement. Objects can be either direct objects or indirect objects.

See Direct Object and Indirect Object.

Parenthetical expression

Parenthetical expressions are expressions which are set off by commas and which seek to add some clarity to a sentence.

Example: "Yogurt, on the other hand, is a fine substitute for ice-cream." "On the other hand" is a parenthetical expression and could be removed from the sentence without destroying sentence meaning.

Examples: *after all, by the way, for example, however, incidentally, indeed, in fact, in my opinion, naturally, nevertheless, of course, on the contrary, on the other hand, to tell you the truth.*

Participle A participle is a verb form (ending in "ed" or "ing") that can function as an adjective. A participle is a type of verbal. Refer to the definition of "verbal."

Examples: "Cars parked near emergency exits will be towed." ("Parked" is a participle; it's an adjective describing "cars." The actual verb in the sentence is "will be towed." "A sleeping dog never bites anyone." The participle "sleeping" describes "dog." The actual verb in the sentence is "bites."

Participle phrase A participle phrase (also called a participial phrase) is a group of related words that contains a participle and, as a unit, typically functions as an adjective.

Examples: "Allowing plenty of time, Bill started studying twelve weeks before taking his College Board exams." ("Allowing plenty of time" functions a participle phrase in describing "Bill.")

Personal pronoun A personal pronoun is a pronoun designating the person speaking, the person spoken to, or the person or thing spoken about.

The following is a complete list of personal pronouns: *I, he, her, him, his, it, its, me, mine, my, our, ours, she, their, theirs, them, they, us, we, who, whom, whose, you, your, yours.*

Phrase A phrase is a group of words which doesn't contain both a subject and a verb. Examples: "Learning to be happy is difficult for a variety of reasons." The phrase "for a variety of reasons" does not contain a verb.

Predicate A predicate is one of the two principal parts of a sentence. The predicate is "any word or words that talk about the subject"; the subject is "the word or words being talked about." Technically, the word "predicate" is a broader term than the word "verb," referring to both a verb and its possible complement. It is, however, much more common to refer to the *verb* and *complement* separately. In such cases, the *verb* can be referred to as the *simple predicate*; the *predicate* is referred to as the *complete predicate*.

Examples: "Water is the key to our survival." In this sentence, the subject is "water" and the predicate is "is the key to our survival." Breaking things down further, the predicate consists of the verb "is" and the complement "the key to our survival."

Reflexive pronoun A reflexive pronoun refers back to a given noun or pronoun.

The following is a complete list of reflexive pronouns: *herself, himself, itself, myself, ourselves, themselves, yourself.*

Relative clause A relative clause is a group of related words that begins with a relative pronoun, and as a unit, functions as an adjective. A relative clause is commonly referred to as an adjective clause (and sometimes as a subordinate adjective clause).

Examples: "Jim Thompson, who mysteriously disappeared while going for an afternoon walk on Easter Sunday, is credited with having revitalized the silk trade in Thailand." "Who mysteriously disappeared while going for an afternoon walk on Easter Sunday" is a relative clause which serves to modify "Jim Thompson."

See also Adjective Clause and Subordinate Clause.

Relative pronoun A relative pronoun modifies a noun or pronoun (called its antecedent). A relative pronoun also begins a relative clause (also known as a subordinate adjective clause).

Examples: There are five relative pronouns in English: *that, which, who, whom,* and *whose.*

Restrictive clause A restrictive clause is essential to the meaning of a sentence. Restrictive clauses are not enclosed by commas.

Example: "The book that is red is the one you need for English class." "That is red" is a restrictive clause.

Run-on sentence A run-on sentence refers to two sentences that are inappropriately joined together, usually by a comma.

Example: "The weather is great, I'm going to the beach." (A comma cannot join two complete sentences. See *Editing II – Punctuation Highlights* for further discussion on how to fix a run-on sentence).

Sentence A sentence is a group of words that contains a subject and a verb, and can stand on its own as a complete thought.

Example: "The world is a stage." The subject is "the world" while the verb is "is"; the complete thought involves comparing the world to a stage.

Sentence fragment A sentence fragment is a group of words that cannot stand on their own to form a complete thought.

Example: "A fine day." This statement is a fragment. It does not constitute a complete thought and cannot stand on its own. The fragment can be turned into a sentence by adding a subject ("today") and a verb ("is")—*Today is a fine day.*

Sentence fragments are not acceptable for use in formal writing. In contrast, sentence fragments are commonly used in informal writing situations (e.g., e-mail and text messaging), and frequently seen in creative communications such as advertising, fiction writing, and poetry.

GRAMMATICAL MUNCHKINS

The following sentence fragments would be acceptable in informal written communication:

Will Michael Phelps' feat of eight Olympic gold medals ever be equaled? Never.

We need to bring education to the world. But how?

Dream on! No one beats Brazil at football when its star forwards show up to play.

Split infinitive A split infinitive occurs when a word (usually an adverb) is placed between the two words that create an infinitive (i.e., between the word "to" and its accompanying verb). Splitting an infinitive is still considered a substandard practice in formal writing.

Example: The sentence, "To boldly go where no one has gone before," contains a split infinitive. The sentence should be rewritten, "To go boldly where no one has gone before."

Subordinate clause A subordinate clause is a clause that cannot stand on its own as a complete sentence. It must instead be combined with at least one independent clause to form a complete sentence. Subordinate clauses are sometimes called dependent clauses.

Example: "We should support the winning candidate whomever that may be." The subordinate clause is "whomever that may be." The independent clause is, "We should support the winning candidate."

Subordinating conjunction

A subordinating conjunction is a conjunction that begins an adverb clause, and serves to join that clause to the rest of the sentence.

Examples: *after, although, as, as if, as long as, as though, because, before, if, in order that, provided that, since, so that, than, though, unless, until, when, whenever, where, wherever, whether, while.*

Note that many of the words in the above list, when used in different contexts, may also function as other parts of speech.

Transitive verb

Transitive verbs require an object to complete its meaning.

Example: "She posted a letter." The verb "posted" requires an object, in this case "letter," to complete its meaning.

See also Intransitive Verb.

Verbal

A verbal is a verb form that functions as a noun, adjective, or adverb. There are three types of verbals: gerunds, infinitives, and participles. Gerunds, infinitives, and participles can form phrases, in which case they are referred to as gerund phrases, infinitive phrases, and participle phrases.

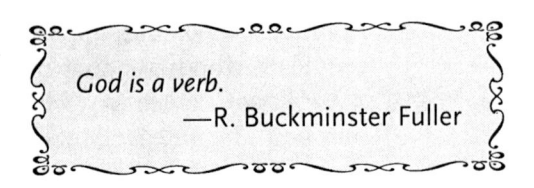

> God is a verb.
> —R. Buckminster Fuller

Chapter 3
Word Gremlins

Diction, sometimes called word usage, is about word choice. A writer often must choose between two similar words or expressions. The good news is that a small amount of concentrated study can greatly improve anyone's grasp of diction. The bad news is that most diction errors will not be picked up by a spell checker or grammar checker. Rather, they lurk in material until discovered and culled.

Idioms, here called grammatical idioms, are words and expressions that have become accepted due to the passage of time. They are right simply because "they're right." Even native speakers have trouble relying entirely on what sounds right. Fortunately, the compiled list of 200 common grammatical idioms will provide a welcome reference for preview and review.

Diction Showdown

Affect, Effect

Affect is a verb meaning "to influence." *Effect* is a noun meaning "result." *Effect* is also a verb meaning "to bring about."

The change in company policy will not <u>affect</u> our pay.

The long-term <u>effect</u> of space travel is not yet known.

A good mentor seeks to <u>effect</u> positive change.

Afterward, Afterwards

These words are interchangeable; *afterward* is more commonly used in America while *afterwards* is more commonly used in Britain. A given document should show consistent treatment.

Allot, Alot, A lot

Allot is a verb meaning "to distribute" or "to apportion." *Alot* is not a word in the English language, but a common misspelling. *A lot* means "many."

To become proficient at yoga one must <u>allot</u> twenty minutes a day to practice.

Having <u>a lot</u> of free time is always a luxury.

All ready, Already

All ready means "entirely ready" or "prepared"; *already* means "before or previously," but may also mean "now or soon."

Contingency plans ensure we are <u>all ready</u> in case the unexpected happens. (entirely ready or prepared)

We've already tried the newest brand. (before or previously)

Is it lunchtime already? (now or so soon)

All together, Altogether

All together means "in one group." *Altogether* has two meanings. It can mean "completely," "wholly," or "entirely." It can also mean "in total."

Those going camping must be all together before they can board the bus.

The recommendation is altogether wrong.

There are six rooms altogether.

NOTE ☙ The phrase "putting it all together" (four words) is correct. It means "putting it all in one place." The phrase "putting it altogether" (three words) is incorrect because it would effectively mean, "putting it completely" or "putting it in total."

Among, Amongst

These words are interchangeable: *among* is American English while *amongst* is British English. A given document should show consistent treatment.

Anymore, Any more

These words are not interchangeable. *Anymore* means "from this point forward"; *any more* refers to an unspecified additional amount.

I'm not going to dwell on this mishap anymore.

Are there any more tickets left?

Anyone, Any one

These words are not interchangeable. *Anyone* means "any person" whereas *any one* means "any single person, item, or thing."

<u>Anyone</u> can take the exam.

<u>Any one</u> of these green vegetables is good for you.

Anytime, Any time

These words are not necessarily interchangeable. *Anytime* is best thought of as an adverb which refers to "an unspecified period of time." *Any time* is an adjective-noun combination which means "an amount of time." Also, *any time* is always written as two words when it is preceded by the preposition "at"; in that case, its meaning is the same as its single word compatriot.

Call me <u>anytime</u> and we'll do lunch.

This weekend, I won't have <u>any time</u> to tweet (twitter).

At <u>any time</u> of the day, you can hear traffic if your window is open.

Anyway, Any way

These words are not interchangeable. *Anyway* means "nevertheless, no matter what the situation is" or "in any case, no matter what." *Any way* means "any method or means."

Keep the printer. I wasn't using it <u>anyway</u>.

Is there <u>any way</u> of salvaging this umbrella?

NOTE ❧ The word "anyways," previously considered non-standard, is now considered an acceptable variant of "anyway," according to the *Merriam-Webster Collegiate Dictionary*.

Apart, A part

These words are not interchangeable. *Apart* means "in separate pieces"; *a part* means "a single piece or component."

Overhaul the machine by first taking it apart.

Every childhood memory is a part of our collective memory.

Awhile, A while

These words are not interchangeable. *Awhile* is an adverb meaning "for a short time"; *a while* is a noun phrase meaning "some time" and is usually preceded by "for."

Let's wait awhile.

I'm going to be gone for a while.

NOTE ❧ It is not correct to write: "Let's wait for awhile."

As, Because, Since

These three words, when used as a conjunction meaning "for the reason that," are all interchangeable.

As everyone knows how to swim, let's go snorkeling.

Because all the youngsters had fishing rods, they went fishing.

Since we have firewood, we'll make a bonfire.

Assure, Ensure, Insure

Assure is to inform positively. *Insure* is to arrange for financial payment in the case of loss. Both *ensure* and *insure* are now largely interchangeable in the sense of "to make certain." *Ensure*, however, implies a kind of virtual guarantee. *Insure* implies the taking of precautionary or preventative measures.

Don't worry. I <u>assure</u> you I'll be there by 8 a.m.

When shipping valuable antiques, a sender must <u>insure</u> any piece for its market value in the event it's damaged or lost.

Hard work is the best way to <u>ensure</u> success regardless of the endeavor.

Every large jewelry shop maintains an on-site safe to <u>insure</u> that inventory is secure during closing hours. (taking of precautionary measures)

Because of, Due to, Owing to

These word pairings are interchangeable and mean "as a result of."

The climate is warming <u>because of</u> fossil fuel emissions.

Fossil fuel emissions are increasing <u>due to</u> industrialization.

<u>Owing to</u> global warming, the weather is less predictable.

Better, Best

Better is used when comparing two things. *Best* is used when comparing three or more things.

Comparing Dan with Joe, Joe is the <u>better</u> cyclist.
Tina is the <u>best</u> student in the class.

Between, Among

Use *between* to discuss two things. Use *among* to discuss three or more things.

The jackpot was divided between two winners.

Five plaintiffs were among the recipients of a cash settlement.

Cannot, Can not

These words are interchangeable with "cannot" being by far the most popular written expression in both American and British English.

Choose, Choosing, Chose, Chosen

Choose is present tense of the verb (rhymes with "blues"). *Choosing* is the present participle (rhymes with "cruising"). *Chose* is the past tense (rhymes with "blows"). *Chosen* is the past participle.

My plan was to choose blue or green for my company logo.

I ended up choosing teal, which is a blend of both colors.

Actually, we first chose turquoise but, soon after, realized that the shade we had chosen was a bit too bright.

NOTE ଔ There is no such word as "chosing." This word is sometimes mistaken for, and incorrectly used in place of, the present participle "choosing."

Complement, Compliment

Both complement or compliment can be used as nouns or verbs. As a verb, *complement* means "to fill in," "to complete," or "to add to and make better"; as a noun it means "something that completes" or "something that improves." *Compliment* is used in two related ways. It is either "an expression of praise" (noun) or is used "to express praise" (verb).

A visit to the Greek islands is a perfect complement to any tour of bustling Athens. Visitors to the Greek island of Mykonos, for instance, are always struck by how the blue ocean complements the white, coastal buildings.

Throughout the awards ceremony, winners and runner-ups received compliments on a job well done. At closing, it was the attendees that complimented the organizers on a terrific event.

Complementary, Complimentary

Both words are used as adjectives. Like complement, *complementary* means "to make complete," "to enhance," or "to improve" (e.g., complementary plans). *Complimentary* means "to praise" (e.g., complementary remarks) or "to receive or supply free of charge."

Only one thing is certain in the world of haute couture: fashion parties brimming with complimentary Champagne and endless banter on how colorful characters and complementary personalities rose to the occasion.

Differs from, Differ with

Use *differ from* in discussing characteristics. Use *differ with* to convey the idea of disagreement.

American English differs from British English.

The clerk <u>differs with</u> her manager on his decision to hire an additional salesperson.

Different from, Different than

These two word pairings are interchangeable. However, whereas *different from* is used to compare two nouns or phases, *different than* is commonly used when what follows is a clause.

Dolphins are <u>different from</u> porpoises.

My old neighborhood is <u>different than</u> it used to be.

Do to, Due to

Do to consists of the verb "do" followed by the preposition "to." *Due to* is an adverbial phrase meaning "because of" or "owing to." *Due to* is sometimes erroneously written as *do to*.

What can we <u>do to</u> save the mountain gorilla?

Roads are slippery <u>due to</u> heavy rain.

Each other, One another

Use *each other* when referring to two people; use *one another* when referring to more than two people.

Two weight lifters helped spot <u>each other</u>.

Olympic athletes compete against <u>one another</u>.

Everyday, Every day

These words are not interchangeable. *Everyday* is an adjective that means either "ordinary" or "unremarkable" (everyday chores) or happening each day (an everyday occurrence). *Every day* is an adverb meaning "each day" or "every single day."

Although we're fond of talking about the everyday person, it's difficult to know what this really means.

Health practicers say we should eat fresh fruit every day.

Everyplace, Every place

These words are not interchangeable. *Everyplace* has the same meaning as "everywhere." *Every place* means in "each space" or "each spot."

We looked everyplace for that DVD.

Every place was taken by the time she arrived.

Everyone, Every one

These words are not interchangeable. *Everyone* means "everybody in a group" whereas *every one* means "each person."

Everyone knows who did it!

Every one of the runners who crossed the finish line was exhausted but jubilant.

Everything, Every thing

Everything means "all things." *Every thing* means "each thing." Note that the word *everything* is much more common than its two-word counterpart.

Everything in this store is on sale.

Just because we don't understand the role that each living organism plays, doesn't mean that to every thing there isn't a purpose.

Every time, Everytime

Every time means "at any and all times." It is always spelled as two separate words. Spelling it as one word is nonstandard and incorrect.

<u>Every time</u> we visit there's always lots of food and drink.

Farther, Further

Use *farther* when referring to distance. Use *further* in all other situations, particularly when referring to extent or degree.

The town is one mile <u>farther</u> along the road.

We must pursue this idea <u>further</u>.

Fewer, Less

Fewer refers to things that can be counted, e.g., people, marbles, accidents. *Less* refers to things that cannot be counted, e.g., money, water, sand.

There are <u>fewer</u> students in class than before the midterm exam.

There is <u>less</u> water in the bucket due to evaporation.

If, Whether

Use *if* to express one possibility, especially conditional statements. Use *whether* to express two (or more) possibilities.

The company claims that you will be successful <u>if</u> you listen to their tapes on motivation.

Success depends on <u>whether</u> one has desire and determination. (The implied "whether or not" creates two possibilities.)

NOTE ☙ In colloquial English, *if* and *whether* are now interchangeable. Either of the following sentences would be correct: "I'm not sure whether I'm going to the party."/"I'm not sure if I'm going to the party."

Instead of, Rather than

These word pairs are considered interchangeable.

Lisa ordered Rocky Road ice cream instead of Mint Chocolate.

The customer wanted a refund rather than an exchange.

Infer, Imply

Infer means "to draw a conclusion"; readers or listeners infer. *Imply* means "to hint" or to suggest"; speakers or writers imply.

I infer from your letter that conditions have improved.

Do you mean to imply that conditions have improved?

Into, In to

These words are not interchangeable. *Into* means "something inside something else"; the phrase *in to* means "something is passing from one place to another."

The last I saw she was walking into the cafeteria.

He finally turned his assignment in to the teacher.

Regarding the sentence above, unless the student were a magician, we could not write, "He finally turned his assignment into the teacher."

Its, It's

Its is a possessive pronoun. *It's* is a contraction for "it is" or "it has."

The world has lost its glory.

It's time to start anew.

Lead, Led

The verb *lead* means "to guide, direct, command, or cause to follow." *Lead* is the present tense of the verb while *led* forms the past tense (and past participle).

More than any other player, the captain is expected to lead his team during the playoffs. Last season, however, it was our goalie, not the captain, who actually led our team to victory.

It is a common mistake to write "lead," when what is called for is "led." This error likely arises given that the irregular verb "read" is spelled the same in the present tense and in the past tense. Ex. "I read the newspaper everyday, and yesterday, I read an amazing story about a 'tree' man whose arms and legs resembled bark."

Lets, Let's

Lets is a verb meaning "to allow or permit." *Let's* is a contraction for "let us."

Technology lets us live more easily.

Let's not forget those who fight for our liberties.

Lie, Lay

In the present tense, *lie* means "to rest" and *lay* means "to put" or "to place." Lie is an intransitive verb (a verb that does not require a direct object to complete its meaning), while lay is a transitive verb (a verb that requires a direct object to complete its meaning).

Lie

Present	Lie on the sofa.
Past	He lay down for an hour.
Perfect Participle	He has lain there for an hour.
Present Participle	It was nearly noon and he was still lying on the sofa.

Lay

Present	Lay the magazine on the table.
Past	She laid the magazine there yesterday.
Perfect Participle	She has laid the magazine there many times.
Present Participle	Laying the magazine on the table, she stood up and left the room.

NOTE ⌘ There is no such word as "layed." This word is the mistaken misspelling of "laid." Ex. "A magazine cover that is professionally laid out," not "a magazine cover that is professionally layed out."

Like, Such as

Such as is used for listing items in a series. *Like* should not be used for listing items in a series. However, *like* is okay to use when introducing a single item.

A beginning rugby player must master many different skills <u>such as</u> running and passing, blocking and tackling, drop kicking, and scrum control.

Dark fruits, <u>like</u> beets, have an especially good cleansing quality.

Loose, Lose, Loss

Loose is an adjective meaning "not firmly attached" or "not tightly drawn." *Lose* is a verb meaning "to suffer a setback or deprivation." *Loss* is a noun meaning "a failure to achieve."

A <u>loose</u> screw will fall out if not tightened.

There is some truth to the idea that if you're going to <u>lose</u>, you might as well <u>lose</u> big.

<u>Loss</u> of habitat is a greater threat to wildlife conservation than is poaching.

Maybe, May be

These words are not interchangeable. *Maybe* is an adverb meaning "perhaps." *May be* is a verb phrase.

<u>Maybe</u> it's time to try again.

It <u>may be</u> necessary to resort to extreme measures.

Might, May

Although *might* and *may* both express a degree of uncertainty, they have somewhat different meanings. *Might* expresses more uncertainty than does *may*. Also, only *might* is the correct choice when referring to past situations.

I <u>might</u> like to visit the Taj Mahal someday. (much uncertainty)

I <u>may</u> go sightseeing this weekend. (less uncertainty)

They <u>might</u> have left a message for us at the hotel. (past situation)

No one, Noone

No one means "no person." It should be spelled as two separate words. The one-word spelling is nonstandard and incorrect.

<u>No one</u> can predict the future.

Number, Amount

Use *number* when speaking of things that can be counted. Use *amount* when speaking of things that cannot be counted.

The <u>number</u> of marbles in the bag is seven.

The <u>amount</u> of topsoil has eroded considerably.

Onto, On to

These words are not equivalent. *Onto* refers to "something placed on something else." The phrase *on to* consists of the adverb "on" and the preposition "to."

Ferry passengers could be seen holding <u>onto</u> the safety rail. We passed the information <u>on to</u> our friends.

Note that we could not pass the information *onto* our friends unless the information was placed physically on top of them.

Passed, Past

Passed functions as a verb. *Past* functions as a noun, adjective, or preposition.

Yesterday, Cindy found out that she passed her much-feared anatomy exam.

The proactive mind does not dwell on events of the past.

Principal, Principle

Although *principal* can refer to the head administrator of a school or even an original amount of money on loan, it is usually used as an adjective meaning "main," "primary," or "most important." *Principle* is used in one of two senses: to refer to a general scientific law or to describe a person's fundamental belief system.

Lack of clearly defined goals is the principal cause of failure.

To be a physicist one must clearly understand the principles of mathematics.

A person of principle lives by a moral code.

Sometime, Some time

These words are not interchangeable. *Sometime* refers to "an unspecified, often longer period of time"; *some time* refers to "a specified, often shorter period of time."

Let's have lunch sometime.

We went fishing early in the morning, but it was some time before we landed our first trout.

Than, Then

Than is a conjunction used in making comparisons. *Then* is an adverb indicating time.

There is controversy over whether the Petronas Towers in Malaysia is taller <u>than</u> the Sears Tower in Chicago.

Finish your work first, <u>then</u> give me a call.

That, Which

The words *which* and *that* mean essentially the same thing. But in context they are used differently. It is common practice to use *which* with nonrestrictive (nonessential) phrases and clauses and to use *that* with restrictive (essential) phrases and clauses. Nonrestrictive phrases are typically enclosed with commas, whereas restrictive phrases are never enclosed with commas. This treatment means that *which* appears in phrases set off by commas whereas *that* does not appear in phrases set off by commas.

The insect <u>that</u> has the shortest lifespan is the Mayfly.

The Mayfly, <u>which</u> lives less than 24 hours, has the shortest lifespan of any insect.

That, Which, Who

In general, *who* is used to refer to people, *which* is used to refer to things, and *that* can refer to either people or things. When referring to people, the choice between *that* and *who* should be based on what feels more natural.

Choose a person <u>that</u> can take charge.

The person <u>who</u> is most likely to succeed is often not an obvious choice.

On occasion, *who* is used to refer to non-persons while *which* may refer to people.

I have a dog <u>who</u> is animated and has a great personality.

<u>Which</u> child won the award? (The pronoun *which* is used to refer to a person.)

There, Their, They're

There is an adverb; *their* is a possessive pronoun; *they're* is a contraction for "they are."

<u>There</u> is a rugby game tonight.

<u>Their</u> new TV has incredibly clear definition.

<u>They're</u> a strange but happy couple.

Toward, Towards

These words are interchangeable: *toward* is American English while *towards* is British English. A given document should show consistent treatment.

Used to, Use to

These words are not interchangeable. *Used to* is the correct form for habitual action. However, when "did" precedes "used to" the correct form is *use to*.

I <u>used to</u> go to the movies all the time.

I didn't <u>use to</u> daydream.

Who, Whom

"Who" is the subjective form of the pronoun and "whom" is the objective form. The following is a good rule in deciding between *who* and *whom:* If "he, she, or they" can be substituted for a pronoun in context, the correct form is *who*. If "him, her, or them" can be substituted for a pronoun in context, the correct form is *whom*. Another very useful rule is that pronouns take their objective forms when they are the direct objects of prepositions.

Let's reward the person who can find the best solution.

Test: "He" or "she" can find the best solution, so the subjective form of the pronoun—who—is correct.

The report was compiled by whom?

Test: This report was drafted by "him" or "her," so the objective form of the pronoun—whom—is correct. Another way of confirming this is to note that "whom" functions as the direct object of the preposition "by," so the objective form of the pronoun is correct.

One particularly tricky situation occurs in the following: "She asked to speak to whoever was on duty." At first glance, it looks as though "whomever" should be correct in so far as "who" appears to be the object of the preposition "to." However, the whole clause "whoever was on duty" is functioning as the direct object of the preposition "to." The key is to analyze the function of "whoever" within the applicable clause itself; in this case, "whoever" is functioning as the subject of the verb "is," thereby taking the subjective form. We can test this by saying *"he* or *she* was on duty."

Let's analyze two more situations, both of which are correct.

1) "I will interview <u>whomever</u> I can find for the job." The important thing is to analyze the role of "whomever" within the clause "whomever I can find" and test it as "I can find *him* or *her*." This confirms that the objective form of the pronoun is correct. In this instance, the whole clause "whomever I can find" is modifying the word form "will interview."

2) "I will give the position to <u>whoever</u> I think is right for the job." Again, the critical thing is to analyze the role of "whoever" within the clause "whoever I think is right for the job." Since we can say "I think he or she is right for the job," this confirms that the subjective form of the pronoun is correct. In this instance, the whole clause "whoever I think is right for the job" is modifying the preposition "to." Therefore, this example mirrors the previous example, "She asked to speak to <u>whoever</u> was on duty."

Whose, Who's

Whose is a possessive pronoun; *who's* is a contraction for "who is."

<u>Whose</u> set of keys did I find?

He is the player <u>who's</u> most likely to make the NBA.

Your, You're

Your is a possessive pronoun; *you're* is a contraction for "you are."

This is <u>your</u> book.

<u>You're</u> becoming the person you want to be.

200 Common Grammatical Idioms

ABC

1. able to X
2. account for
3. according to
4. a craving for
5. a debate over
6. a descendant of
7. affiliated with
8. agree to (a plan or action)
9. agree with (person/idea)
10. allow(s) for
11. amount to
12. a native of
13. angry at/angry with
14. appeal to
15. apply to/apply for
16. approve(d) of/disapprove(d) of
17. a responsibility to
18. argue with/over
19. a sequence of
20. as a consequence of X
21. as...as
22. as...as do/as...as does
23. as a result of
24. as good as
25. as good as or better than
26. as great as
27. as many X as Y
28. as much as
29. as X is to Y
30. ask X to do Y
31. associate with
32. attempt to
33. attend to
34. attest to

35. attribute X to Y
36. assure that
37. averse to
38. based on
39. be afraid of
40. because of
41. believe X to be Y
42. better served by X than by Y
43. better than
44. between X and Y
45. both X and Y
46. capable of
47. centers on
48. choice of
49. choose from/choose to
50. claim to be
51. collaborate with
52. compare to/compare with
53. comply with
54. composed of
55. concerned about/with (not "concerned at")
56. conform to
57. conclude that
58. connection between X and Y
59. consider(ed) (without "to be")
60. consider(ed) (with "to be")
61. consistent with
62. contend that
63. contrast X with Y
64. convert to
65. cost of/cost to
66. credit(ed) X with having

DEF

67. debate over
68. decide on/decide to
69. declare X to be Y

70.	defend against
71.	define(d) as
72.	delighted by
73.	demand that
74.	demonstrate that
75.	depend(ent) on
76.	depends on whether
77.	depict(ed) as
78.	descend(ed) from
79.	desirous of
80.	determined by
81.	differ from/differ with
82.	different from
83.	difficult to
84.	disagree with (person/idea)
85.	discourage from
86.	differentiate between X and Y
87.	differentiate X from Y
88.	dispute whether
89.	distinguish X from Y
90.	divergent from
91.	do so/doing so (not "do it"/"doing it")
92.	doubt that
93.	draw on
94.	either X or Y
95.	enable X to Y
96.	enamored of/with
97.	enough X that Y
98.	estimated to be
99.	expect to
100.	expose(d) to
101.	fascinated by
102.	fluctuations in
103.	forbid X and Y
104.	frequency of
105.	from X rather than from Y (not "from X instead of Y")
106.	from X to Y

GHI

107. give credit for/give credit to
108. hypothesize that
109. in an effort to
110. in association with
111. indifferent toward(s)
112. infected with
113. inherit X from Y
114. in order to
115. in reference to/with reference to
116. in regard to/with regard to
117. in search of
118. insists that
119. intend(ed) to
120. intersection of X and Y
121. in the same way as ... to
122. in the same way that
123. introduce(d) to
124. in violation of
125. isolate(d) from

JKL

126. just as X, so (too) Y
127. less X than Y
128. likely to/likely to be
129. liken to

MNO

130. meet with
131. mistake (mistook) X for Y
132. model(ed) after
133. more common among X than among Y
134. more ... than ever
135. more X than Y
136. native to

137. neither X nor Y
138. no less...than
139. no less was X than was Y
140. not X but rather Y
141. not only X but (also) Y
142. not so much X as Y
143. on account of
144. on the one hand/on the other hand

PQR

145. opposed to/opposition to
146. opposite of
147. inclined to
148. in comparison to
149. in conjunction with
150. in contrast to
151. in danger of
152. independent from
153. owing to
154. persuade X to Y
155. partake (partook) of
156. permit X to Y
157. potential to
158. prefer X to Y (not "prefer X over Y")
159. preferable to
160. prejudiced against
161. prevent from
162. prized by
163. prohibit X from Y
164. protect against
165. question whether
166. range(s) from X to Y
167. rates for (not "rates of")
168. recover from X
169. recover X from Y
170. regard(ed) as
171. replace(d) with

172. responsible for
173. resulting in

STU

174. sacrifice X for Y
175. seem to indicate
176. similar to
177. so as not to be hindered by
178. so X as to be Y
179. so X as to constitute Y
180. so X that Y
181. subscribe to
182. such X as Y and Z
183. sympathy for
184. sympathize with
185. tamper with
186. targeted at
187. the more X the greater Y
188. the same to X as to Y
189. to result in
190. to think of X as Y
191. tying X to Y
192. used to (not "use to")

VWXZY

193. view X as Y
194. whether X or Y
195. worry about (not "over")
196. X enough to Y
197. X instead of Y
198. X is attributed to Y
199. X out of Y (numbers)
200. X regarded as Y

Chapter 4
Putting It All Together

These thirty all-star grammar problems are grouped according to the "big six" grammar categories as introduced in the *100-Question Quiz* (chapter 1). Although grammar is the driving force behind each problem, issues relating to diction or idioms may be interwoven subcomponents. Read each question, deciding whether the underlined portion of each sentence needs to be changed to yield the *best* answer. If you feel the original sentence reads best among the choices, choose choice A, which is merely a restatement of the original; otherwise choose one of the variant answers B through E.

In addition to the answers and explanations included at the end of this section, each problem has a *Skill Rating*—easy, medium, or difficult—as well as a *Snapshot* caption to briefly explain why a given problem was chosen.

30 All-Star Grammar Problems

Subject-Verb Agreement:

1. Vacation

<u>Neither Martha or her sisters are going on vacation.</u>

A) Neither Martha or her sisters are going on vacation.

B) Neither Martha or her sisters is going on vacation.

C) Neither any of her sisters nor Martha are going on vacation.

D) Neither Martha nor her sisters are going on vacation.

E) Neither Martha nor her sisters is going on vacation.

2. Leader

The activities of our current leader <u>have led to a significant increase in the number of issues relating to the role of the military in non-military, nation-building exercises.</u>

A) have led to a significant increase in the number of issues relating to the role of the military in non-military, nation-building exercises.

B) have been significant in the increase in the amount of issues relating to the role of the military in non-military, nation-building exercises.

C) has led to a significant increase in the number of issues relating to the role of the military in non-military, nation-building exercises.

PUTTING IT ALL TOGETHER

D) has been significant in the increase in the number of issues relating to the role of the military in non-military, nation-building exercises.

E) has significantly increased the amount of issues relating to the role of the military in non-military, nation-building exercises.

3. Marsupial

According to scientists at the University of California, the pattern of changes that have occurred in placental DNA over the millennia <u>indicate the possibility that every marsupial alive today might be descended from a single female ancestor that</u> lived in Africa sometime between 125 and 150 million years ago.

A) indicate the possibility that every marsupial alive today might be descended from a single female ancestor that

B) indicate that every marsupial alive today might possibly be a descendant of a single female ancestor that had

C) may indicate that every marsupial alive today has descended from a single female ancestor that had

D) indicates that every marsupial alive today might be a descendant of a single female ancestor that

E) indicates that every marsupial alive today may be a descendant from a single female ancestor that

4. Critics' Choice

<u>In this critically acclaimed film, there are a well-developed plot and an excellent cast of characters.</u>

A) In this critically acclaimed film, there are a well-developed plot and an excellent cast of characters.

B) In this critically acclaimed film, there is a well-developed plot and an excellent cast of characters.

C) In this film, which is critically acclaimed, there is a well-developed plot and an excellent cast of characters.

D) In this film, which has been critically acclaimed, there are a well-developed plot and an excellent cast of characters.

E) There is a well-developed plot and an excellent cast of characters in this critically acclaimed film.

5. Recommendations

<u>Implementing the consultants' recommendations is expected to result in</u> both increased productivity and decreased costs.

A) Implementing the consultants' recommendations is expected to result in

B) Implementing the consultants' recommendations are expected to result in

C) The expected result of enacting the consultants' recommendations are

D) The expected results of enacting the consultants' recommendations is

E) It is expected that enactment of the consultants' recommendations are to result in

PUTTING IT ALL TOGETHER

Pronoun Usage:

6. Valuation

Financial formulas for valuing companies do not apply to Internet companies in the same way as they do to traditional businesses, because they are growing and seldom have ascertainable sales and cash flows.

A) Financial formulas for valuing companies do not apply to Internet companies in the same way as they do to traditional businesses, because they are growing and seldom have ascertainable sales and cash flows.

B) Internet companies are not subject to the same applicability of financial formulas for valuing these companies as compared with traditional businesses, because they are growing and seldom have ascertainable sales and cash flows.

C) Because they are growing and seldom have ascertainable sales and cash flows, financial formulas for valuing companies do not apply to Internet companies in the same way as they do to traditional businesses.

D) Because they are growing and seldom have ascertainable sales and cash flows, Internet companies are not subject to the same applicability of financial valuation formulas as are traditional businesses.

E) Because Internet companies are growing and seldom have ascertainable sales and cash flows, financial formulas for valuing these companies do not apply to them in the same way as to traditional businesses.

7. Inland Taipan

The Inland Taipan or Fierce Snake of central Australia is widely <u>regarded to be the world's most venomous snake; the poison from its bite can kill human victims unless treated</u> within thirty minutes of an incident.

A) regarded to be the world's most venomous snake; the poison from its bite can kill human victims unless treated

B) regarded as the world's most venomous snake; the poison from its bite can kill human victims unless treated

C) regarded to be the world's most venomous snake; the poison from its bite can kill human victims unless it is treated

D) regarded as the world's most venomous snake; the poison from its bite can kill human victims unless they are treated

E) regarded to be the world's most venomous snake; the poison from its bite can kill human victims unless they are treated

8. Medicare

<u>Although Medicare legislation is being considered by the House of Representatives, they do not expect it to pass without being significantly revised.</u>

A) Although Medicare legislation is being considered by the House of Representatives, they do not expect it to pass without being significantly revised.

B) Although the House of Representatives is considering Medicare legislation, they do not expect it to pass without significant revision.

C) Although the House of Representatives is considering Medicare legislation, it is not expected to pass without being significantly revised.

D) If it is to be passed, the House of Representatives must significantly revise Medicare legislation.

E) Consideration and significant revision is expected if Medicare legislation is to be passed by the House of Representatives.

9. Oceans

One cannot gauge the immensity of the world's oceans until you have tried to sail around the world.

A) One cannot gauge the immensity of the world's oceans until you have tried to sail around the world.

B) One cannot gauge the immensity of the world's oceans until they have tried to sail around the world.

C) One cannot gauge the immensity of the world's oceans until he or she has tried to sail around the world.

D) A person cannot gauge the immensity of the world's oceans until you have tried to sail around the world.

E) A person cannot gauge the immensity of the world's oceans until they have tried to sail around the world.

Modification:

10. Metal Detector

<u>Using a metal detector, old coins and other valuables can be located by hobbyists even though they are buried in the sand and dirt.</u>

A) Using a metal detector, old coins and other valuables can be located by hobbyists even though they are buried in the sand and dirt.

B) Old coins and other valuables can be located by hobbyists using a metal detector even though they are buried in the sand and dirt.

C) Using a metal detector, hobbyists can locate old coins and other valuables even though they are buried in the sand and dirt.

D) Buried in the sand and dirt, old coins and other valuables can be located by hobbyists using a metal detector.

E) A metal detector can be used to locate old coins and other valuables that are buried in the sand and dirt by a hobbyist.

11. Hungary

<u>With</u> less than one percent of the world's population, Hungarians have made disproportionately large contributions to the fields of modern math and applied science.

A) With

B) Having

C) Despite having

D) Although constituting

E) In addition to accounting for

12. Natural Beauty

Plastic surgeons who perform surgery for non-medical reasons defend their practice on the basis of the free rights of their patients; many others in the health field, however, contend that plastic surgery degrades natural beauty, <u>which they liken to reconstructing a national park.</u>

A) which they liken to reconstructing a national park.

B) which they liken to a national park with reconstruction done to it.

C) which they liken to reconstruction done on a national park.

D) likening it to a national park with reconstruction done to it.

E) likening it to reconstructing a national park.

Parallelism:

13. Cannelloni

Cannelloni has and always will be my favorite Italian dish.

A) Cannelloni has and always will be my favorite Italian dish.

B) Cannelloni was, has, and always will be my favorite Italian dish.

C) Cannelloni was and always will be my favorite Italian dish.

D) Cannelloni has been and always will be my favorite Italian dish.

E) Cannelloni is, has, and always will be my favorite Italian dish.

14. Massage

Massage creates a relaxing, therapeutic, and rejuvenating experience both for your body and your well-being.

A) both for your body and your well-being.

B) for both your body and your well-being.

C) both for your body and well-being.

D) for both your body and well-being.

E) both for your body as well as your well-being.

PUTTING IT ALL TOGETHER

15. Europeans

<u>Italy is famous for its composers and musicians, France, for its chefs and philosophers, and Poland, for its mathematicians and logicians.</u>

A) Italy is famous for its composers and musicians, France, for its chefs and philosophers, and Poland, for its mathematicians and logicians.

B) Italy is famous for its composers and musicians, France for its chefs and philosophers, Poland for its mathematicians and logicians.

C) Italy is famous for its composers and musicians. France for its chefs and philosophers. Poland for its mathematicians and logicians.

D) Italy is famous for their composers and musicians; France, for their chefs and philosophers; Poland for their mathematicians and logicians.

E) Italy, France, and Poland are famous for their composers and musicians, chefs and philosophers, and mathematicians and logicians.

Comparisons:

16. Sweater

Although neither sweater is really the right size, <u>the smallest one fits best.</u>

A) the smallest one fits best.

B) the smallest one fits better.

C) the smallest one is better fitting.

D) the smaller of the two fits best.

E) the smaller one fits better.

17. Sir Isaac Newton

Within the scientific community, the accomplishments of Sir Isaac Newton are referred to more often <u>than any</u> scientist, living or dead.

A) than any

B) than any other

C) than those of any

D) than are those of any

E) than those of any other

18. Soya

In addition to having more protein than meat does, <u>the protein in soybeans is higher in quality than that in meat.</u>

A) the protein in soybeans is higher in quality than that in meat.

B) the protein in soybeans is higher in quality than it is in meat.

C) Soybeans have protein of higher quality than that in meat.

D) Soybean protein is higher in quality than it is in meat.

E) Soybeans have protein higher in quality than meat.

19. Angel

<u>She sings like an angel sings.</u>

A) She sings like an angel sings.

B) She sings like an angel does.

C) She sings as an angel sings.

D) She sings as if an angel.

E) She sings as if like an angel.

20. Perceptions

Because right-brained individuals do not employ convergent thinking processes, <u>like left-brained individuals,</u> they may not notice and remember the same level of detail as their counterparts.

A) like left-brained individuals,

B) unlike a left-brained individual,

C) as left-brained individuals,

D) as left-brained individuals do,

E) as a left-brained individual can,

21. Geography

Despite the fact that the United States is a superpower, <u>American high school students perform more poorly on tests of world geography and international affairs than do</u> their Canadian counterparts.

A) American high school students perform more poorly on tests of world geography and international affairs than do

B) American high school students perform more poorly on tests of world geography and international affairs as compared with

C) American high school students perform more poorly on tests of world geography and international affairs as compared to

D) the American high school student performs more poorly on tests of world geography and international affairs than does

E) the American high school student performs more poorly on tests of world geography and international affairs as compared with

22. Assemblée Nationale

<u>As Parliament is the legislative government body of Great Britain,</u> the Assemblée Nationale is the legislative government body of France.

A) As Parliament is the legislative government body of Great Britain,

B) As the legislative government body of Great Britain is Parliament,

C) Just like the legislative government body of Great Britain, which is Parliament,

D) Just as Parliament is the legislative government body of Great Britain, so

E) Just as the government of Britain's legislative branch is Parliament,

23. Bear

<u>Like the Alaskan brown bear and most other members of the bear family, the diet of the grizzly bear consists of both meat and vegetation.</u>

A) Like the Alaskan brown bear and most other members of the bear family, the diet of the grizzly bear consists

B) Like those of the Alaskan brown bear and most other members of the bear family, the diets of a grizzly bear consist

C) Like the Alaskan brown bear and most other members of the bear family, the grizzly bear has a diet consisting

D) Just like the diet of the Alaskan brown bear and most other members of the bear family, the diets of the grizzly bear consist

E) Similar to the diets of the Alaskan brown bear and most other members of the bear family, grizzly bears have a diet which consists

24. Smarts

Unlike the Miller Analogies Test, which follows a standardized format, <u>the formats for IQ tests vary considerably in both content and length.</u>

A) the formats for IQ tests vary considerably in both content and length.

B) the format for an IQ test varies considerably in both content and length.

C) an IQ test follows a format that varies considerably in both content and length.

D) an IQ test follows formats that vary considerably in both content and length.

E) IQ tests follow formats that vary considerably in both content and length.

Verb Tenses:

25. Golden Years

According to the findings of a recent study, many executives <u>had elected early retirement rather than face</u> the threats of job cuts and diminishing retirement benefits.

A) had elected early retirement rather than face

B) had elected to retire early rather than face

C) have elected early retirement instead of facing

D) have elected early retirement rather than facing

E) have elected to retire early rather than face

26. Politics

Although he <u>disapproved of the political platform set forth by Senator Barack Obama during the 2008 U.S. presidential primaries, Senator John McCain had later conceded</u> that there must be a basis for a coalition government and urged members of both parties to seek compromise.

A) disapproved of the political platform set forth by Senator Barack Obama during the 2008 U.S. presidential primaries, Senator John McCain had later conceded

B) has disapproved of the political platform set forth by Senator Barack Obama during the 2008 U.S. presidential primaries, Senator John McCain had later conceded

C) has disapproved of the political platform set forth by Senator Barack Obama during the 2008 U.S. presidential primaries, Senator John McCain later conceded

D) had disapproved of the political platform set forth by Senator Barack Obama during the 2008 U.S. presidential primaries, Senator John McCain later conceded

E) had disapproved of the political platform set forth by Senator Barack Obama during the 2008 U.S. presidential primaries, Senator John McCain had later conceded

27. Trend

The percentage of people remaining single in Holland increased abruptly between 1980 and 1990 and continued to rise more gradually over the next 10 years.

A) The percentage of people remaining single in Holland increased abruptly between 1980 and 1990 and continued to rise more gradually over the next ten years.

B) The percentage of people remaining single in Holland increased abruptly between 1980 and 1990 and has continued to rise more gradually over the next ten years.

C) The percentage of people remaining single in Holland increased abruptly between 1980 and 1990 and had continued to rise more gradually over the next ten years.

D) There had been an abrupt increase in the percentage of people remaining single in Holland between 1980 and 1990 and it continued to rise more gradually over the next ten years.

E) There was an abrupt increase in the percentage of people remaining single in Holland between 1980 and 1990 which continued to rise more gradually over the next ten years.

28. Fire

<u>Most houses that were destroyed and heavily damaged in residential fires last year were</u> built without adequate fire detection apparatus.

A) Most houses that were destroyed and heavily damaged in residential fires last year were

B) Most houses that were destroyed or heavily damaged in residential fires last year had been

C) Most houses that were destroyed and heavily damaged in residential fires last year had been

D) Most houses that were destroyed or heavily damaged in residential fires last year have been

E) Most houses that were destroyed and heavily damaged in residential fires last year have been

29. B-School

<u>As graduate management programs become more competitive in the coming years in terms of their promotional and financial undertakings, schools have been becoming</u> more and more dependent on alumni networks, corporate sponsorships, and philanthropists.

A) As graduate management programs become more competitive in the coming years in terms of their promotional and financial undertakings, schools have been becoming

B) As graduate management programs are becoming more competitive in the coming years in terms of their promotional and financial undertakings, schools have been becoming

C) As graduate management programs become more competitive in the coming years in terms of their promotional and financial undertakings, schools have become

D) As graduate management programs are becoming more competitive in the coming years in terms of their promotional and financial undertakings, schools have become

E) As graduate management programs become more competitive in the coming years in terms of their promotional and financial undertakings, schools will become

30. Summer in Europe

By the time we have reached France, we will have been backpacking for twelve weeks.

A) By the time we have reached France, we will have been backpacking for twelve weeks.

B) By the time we have reached France, we will have backpacked for twelve weeks.

C) By the time we reach France, we will have been backpacking for twelve weeks.

D) By the time we will have reached France, we will have backpacked for twelve weeks.

E) By the time we reached France, we will have been backpacking for twelve weeks.

Answers and Explanations

1. Vacation

Choice D
Classification: Subject-Verb Agreement
Skill Rating: Easy
Snapshot: This problem is included to highlight the handling of correlative conjunctions such as "either/or" and "neither/nor" which may involve the use of a singular or plural verb.

The consistent appearance of "neither" indicates a "neither...nor" relationship. We can eliminate choices A and B outright. The correct verb is said to match what comes after the word "nor." Since "her sisters" in D is plural, the plural verb "are" does the trick.

In summary, singular subjects following "or" or "nor" always take a singular verb; plural subjects following "or" or "nor" take a plural verb. Stated another way, when two items are connected by "or" or "nor," the verb agrees with the closer subject. That is, the verb only needs to agree with the subject that comes after "or" or "nor."

There are two potentially correct answers:

> Neither Martha nor her sisters <u>are</u> going on vacation.
> or
> Neither her sisters nor Martha <u>is</u> going on vacation.

Note that only the first alternative above is presented by answer choice D.

2. Leader

Choice A
Classification: Subject-Verb Agreement
Skill Rating: Easy
Snapshot: This problem is included to show subject-verb agreement and to highlight the role of prepositional phrases in disguising the subject and verb.

The subject of a sentence determines the verb (i.e., singular subjects take singular verbs; plural subjects take plural verbs) and the subject of this sentence is "activities" (plural). The intervening phrase "of our current leader" is a prepositional phrase, and prepositional phrases can never contain the subject of a sentence. Mentally cut out this phrase. Since the subject is "activities," the verb is "have," not "has." Another distinction that needs to be drawn relates to the difference between "number" and "amount." The word "number" is used for countable items and "amount" for non-countable items. Therefore, we have no problem choosing choice A as the correct answer after applying only two rules—the first is a subject-verb agreement rule followed by the "number" versus "amount" semantic distinction. Also, per choices B and D, the clause "has/have been significant in the increase" is not only awkward but also passive.

3. Marsupial

Choice D
Classification: Subject-Verb Agreement
Skill Rating: Medium
Snapshot: This problem is also included to highlight the role of prepositional phrases within subject-verb agreement.

The subject of the sentence is "pattern" which is singular and a singular subject takes the singular verb "indicates." An additional way to eliminate choices A and B is through the redundant use of the words "might" and "possibility" which express the same idea; either "possibility" or "might" is required. Also, the use of "might" in choice D is better than "may" (choice E) because "might" more clearly indicates "possibility" than does "may." In choosing between choices D and E, the idiom "descendant of" is superior to the unidiomatic "descendant from." Finally, note that in choices B and C, "had," the auxiliary of "lived," should be deleted because the simple past tense is correct. The past perfect, which employs "had," is not required; the past perfect tense is used to refer to an action that precedes some other action also occurring in the past.

NOTE ○☙ This problem complements the previous one. The former problem contained a plural subject ("activities") and a single item in the prepositional phrase ("current leader"). This problem contains a singular subject ("pattern") and a plural item in the prepositional phrase ("changes").

4. Critics' Choice

Choice A
Classification: Subject-Verb Agreement
Skill Rating: Medium
Snapshot: This problem is included to highlight "there is/there are" constructions in which the subject of the sentence comes after, not before the verb.

The compound subject is plural—"well-developed plot <u>and</u> an excellent cast of characters"—and therefore requires the plural verb "are." Choices B, C, and E are out because of the incorrect verb "is." Choices C and D employ roundabout constructions which are inferior to "In this critically acclaimed film." Choice D also employs the passive construction "which has been critically acclaimed." Choice E rearranges the sentence but still incorrectly employs the singular verb "is."

5. Recommendations

Choice A
Classification: Subject-Verb Agreement
Skill Rating: Medium
Snapshot: This problem is included to highlight gerund phrases, which, when acting as the subject of a sentence, are always singular.

The gerund phrase "Implementing the consultants' recommendations" is the subject of the sentence. As gerund phrases are always singular, the correct verb here is "is." In choice C, "expected result" requires the verb "is," whereas in choice D, "expected results" requires the verb "are." In choice E, the "it is" construction creates an unnecessarily weak opener and an awkward sentence style.

6. Valuation

Choice E
Classification: Pronoun Usage
Skill Rating: Difficult
Snapshot: This problem is included to highlight ambiguity arising from the use of personal pronouns, and seeks to clear up such ambiguity, not by replacing pronouns, but by rearranging the sentence itself. Part of the reason it garners a high difficulty rating is because the problem is long, and somewhat more difficult to read and analyze.

The problem with choices A and B is that the word "they" refers to traditional businesses; this is illogical because traditional businesses are not growing, Internet companies are. Remember that a pronoun modifies the closest noun that precedes it. The structure in choice C makes it seem as if "financial formulas" are growing, and this, of course, is farcical.

Choices A and C use the awkward, "do not apply to X in the same way as they do to Y." A more succinct rendition is found in choice

E—"do not apply to X in the same way as to Y." In choices A, C, and E, the verb "apply" is more powerful and therefore superior to the noun form "applicability" (as used in choices B and D).

NOTE ☙ Beware of the high school "tall tale" that suggests you shouldn't begin a sentence with the word "because." If you learned this as a rule, forget it. According to the conventions of Standard Written English (SWE)—which, incidentally, this book abides by—the word "because" functions as a subordinating conjunction. Its use is effectively identical to that of "as" or "since," and we can think of these three words as substitutes. In short, there's actually no rule of grammar or style preventing us from beginning a sentence with the word "because."

7. Inland Taipan

Choice D
Classification: Pronoun Usage
Skill Rating: Medium
Snapshot: This problem is included to highlight the occasional need to add pronouns in order to remove ambiguity.

This form of ambiguous reference is subtle. The original sentence is missing "they," and without the pronoun, *they*, the word "treated" might refer to "poison" or "victims"; "treated" is only supposed to refer to "victims." In choice C, the pronoun "it" logically but incorrectly refers to "bite." Technically it is not the bite that needs to be treated but the actual victims. Choices A, C, and E erroneously employ the idiom "regarded to be" when the correct idiom is "regarded as."

8. Medicare

Choice C
Classification: Pronoun
Skill Rating: Easy
Snapshot: This problem is included to highlight the need to choose the correct pronoun—*it*—when referring to a collective singular noun or single inanimate object.

Choices A and B are incorrect because the pronoun "they" cannot refer to the House of Representatives. Not only is the House of Representatives a collective singular noun, but it is also an inanimate object; therefore the proper pronoun choice is "it."

Choice D improperly employs the pronoun "it," which incorrectly refers to the House of Representatives rather than to Medicare legislation. Choice E may be the most passive of these sentences, in which the doer of the action, the House of Representatives, is now at the very back of the sentence.

In choice C, the pronoun "it" correctly refers to Medicare legislation. The subordinate clause "although the House of Representatives is considering Medicare legislation" is written in the active voice. The latter part of the sentence is written in the passive voice "without being significantly revised," and we just have to live with it given that it's the best of the remaining choices. For the record, two alternative wordings for the latter part of the sentence might include: *it is not expected to pass unless it is significantly revised* (active voice but employs two uses of the pronoun "it"); *it is not expected to pass without significant revision* (active voice but employs the nominalized "revision").

NOTE ಐ In general, the five most common signals of the passive voice include: "be," "was," "were," "been," "being." In addition, the preposition "by" is also closely associated with the passive voice: e.g., "The ball was caught by the outfielder."

9. Oceans

Choice C
Classification: Pronoun Usage
Skill Rating: Easy
Snapshot: This problem is included to highlight an improper shift in voice, also known as a shift in point of view.

In choice A, the third-person singular pronoun "one" is improperly matched with the second-person pronoun "you." In choice B, the third-person singular "one" is improperly matched with the third-person plural "they." Pronouns must agree with their antecedents in number and person. The problem highlighted here is not that of agreement with respect to person, but number. In choice C, the third-person singular "one" is properly matched with third-person singular "he" or "she." Per choice D, the third-person singular noun "a person" is improperly matched with the second person "you." Per choice E, the third-person singular noun "a person" is improperly matched with the third-person plural pronoun "they." Pronouns must agree with their antecedents in number and person. The problem here is not agreement with respect to person, but number.

NOTE ଊ The following summarizes the do's and don'ts with respect to pronoun usage in terms of person and number:

1) *You* can only be matched with *you*.

Only "you" goes with "you." After all, there is only one second-person pronoun—*you*.

2) *He, she, one,* or *a person* can be matched with any one of *he, she, one,* or *a person*.

Any third-person singular pronoun (e.g., he, she, one) or third-person singular noun (e.g., a person) can be matched with another third-person pronoun or noun (notwithstanding that gender should match as well).

3. *You* cannot be matched with *he, she, one,* or *a person*.

The second person pronoun "you" does not match any third-person pronoun or noun.

4. *They* cannot be matched with *he, she, one,* or *a person*.

The third-person plural pronoun "they" does not match properly any third-person singular pronoun or noun. Note also that "a person" is a noun, not a pronoun.

10. Metal Detector

Choice C
Classification: Modification
Skill Rating: Easy
Snapshot: This problem is included to illustrate misplaced modifiers. In particular, an introductory modifying phrase (a phrase that begins the sentence) always modifies the first noun or pronoun that follows it (and which itself is in the subjective case). As a general rule: *Modifying words or phrases should be kept close to the word(s) that they modify.*

The only answer choice that is written in the active voice is choice C. The other four answer choices are written in the passive voice (note the word "be," which signals the passive voice). In choice A, coins and other valuables cannot *use* a metal detector; we must look for a person to act as the doer of the action. Choice E changes the meaning of the sentence, suggesting that the hobbyists bury the coins themselves. Whereas choices A and E are incorrect, choices B, C, and D are each grammatically correct. Choice C is the winner because, all things being equal, the active voice is deemed superior to the passive voice. This is a rule of style rather than grammar. Style is more or less effective, better or worse. Grammar is correct or incorrect, right or wrong.

NOTE ଓ Modification may involve the replacement of individual qualifying words, such as *almost, only, just, even, hardly, nearly, not,* and *merely*. Ideally, these words should be placed immediately before the words they modify lest they cause confusion.

In the classic example below, consider how the placement of the word "only," within a sentence, can change the meaning of that sentence.

Original		Life exists on earth.

Let's add the word "only" and vary its placement:

Example 1		Only life exists on earth.

		The meaning is that life is the sole occupier of earth. However, we know that there are things besides life that exist on earth, including inanimate objects like rocks.

Example 2		Life only exists on earth.

		The meaning is that life merely exists on earth and doesn't do anything else.

Example 3		Life exists only on earth.

		This is likely the intended meaning. The word "only" is appropriately placed in front of the word phrase it modifies—*on earth*.

Example 4		Life exists on only earth.

		The meaning here is the same as above but slightly more dramatic. The implication is that life's sole domain is earth, and we're proud of it.

THE LITTLE GOLD GRAMMAR BOOK

Example 5   Life exists on earth only.

The meaning is also the same as example 3, but with a flair for the dramatic. The implication may be that life is found only on earth, and isn't that a shame.

11. Hungary

Choice D
Classification: Modification
Skill Rating: Medium
Snapshot: This problem is included to highlight a modification subtlety which necessitates the use of "account for" or "constitute."

Technically speaking, Hungarians don't *have* less than one percent of the world's population; they "account for" or "constitute" less than one percent of the world's population. This latter option is represented in choice D. The logic of choice E makes it incorrect. The transition words "in addition" are illogical because the sentence construction requires contrast, and the word "although" is consistent in this respect. For the record, another correct answer would have included: "Although accounting for less than one percent of the world's population, Hungarians have made disproportionately large contributions to the fields of modern math and applied sciences."

12. Natural Beauty

Choice E
Classification: Modification
Skill Rating: Medium
Snapshot: This problem is included to highlight another type of modification problem, known as "back sentence modification."

The final answer proves best—correct, logical, and succinct—in comparing *plastic surgery* to the act of *reconstructing a national park*. In short, the patient is being compared to a national park

PUTTING IT ALL TOGETHER

while the act of plastic surgery is being likened to the act of reconstructing a national park. The word "likening" functions as a participle; it introduces the participle phrase "likening it to reconstructing a national park." This phrase properly refers to "surgery," not "natural beauty."

In choices A, B, and C, the relative pronoun "which" refers, not to plastic surgery, but to the noun immediately preceding it, "(natural) beauty." As a result, natural beauty is compared to "reconstructing a national park" (choice A), to "a national park" (choice B), and to "reconstruction" (choice C). Choice D corrects this problem by eliminating the "which" construction and supplying the pronoun "it," thus referring clearly to "plastic surgery," but it illogically compares "plastic surgery" to "a national park." Moreover, the double use of "it" is awkward.

13. Cannelloni

Choice D
Classification: Parallelism
Skill Rating: Easy
Snapshot: This problem is included to highlight the use of parallelism in contexts of ellipsis.

To test choice D, simply complete each component idea, making sure each makes sense. "Cannelloni <u>has been</u> my favorite dish... Cannelloni always <u>will be</u> my favorite dish." Now check this against the original: "Cannelloni <u>was</u> my favorite dish (doesn't work)... Cannelloni always <u>will be</u> my favorite dish." Choice E suffers the same fate as choices A and B, erroneously omitting *has been*. Choices B and C are muddled; the word "was" illogically suggests that Cannelloni was once a favorite dish but no longer is.

THE LITTLE GOLD GRAMMAR BOOK

14. Massage

Choice B
Classification: Parallelism
Skill Rating: Medium
Snapshot: This problem is included to highlight the use of parallelism in correlative conjunctions.

There are six correlative conjunctions in English. These include: *either...or, neither...nor, not only...but (also), both...and, whether...or,* and *just as...so (too).* The purpose of correlative conjunctions is to join ideas of equal weight. Therefore, things on both sides of each connector should be parallel in form and equal in weight.

The word pairing "both...as well as" is unidiomatic, so choice E can be eliminated. Here the correlative conjunction is "both...and." The words that follow "both" and "and" must be parallel in structure. In choice B, the correct answer, the words "your body" and "your well-being" follow on both sides of "both" and "and" in perfect parallelism. Choices C and D are not parallel. For the record, there are effectively two possibilities:

Massage creates a relaxing, therapeutic, and rejuvenating experience for <u>both</u> your body <u>and</u> your well-being.
or
Massage creates a relaxing, therapeutic, and rejuvenating experience <u>both</u> for your body <u>and</u> for your well-being.

Here's another example:

Incorrect	Jonathan <u>not only</u> likes rugby <u>but also</u> kayaking.
Correct	Jonathan likes <u>not only</u> rugby <u>but also</u> kayaking.
or	
Correct	Jonathan <u>not only</u> likes rugby <u>but also</u> likes kayaking.

Two more correct versions would include:

Incorrect	Sheila <u>both</u> likes to act <u>and</u> to sing.
Correct	Sheila likes <u>both</u> to act <u>and</u> to sing.
or	
Correct	Sheila <u>both</u> likes to act <u>and</u> likes to sing.

15. Europeans

Choice A
Classification: Parallelism
Skill Rating: Difficult
Snapshot: This problem is included to highlight the use of parallelism with regard to ellipsis, and to review semicolons, omission commas, sentence run-ons and sentence fragments.

In choice A, the comma placed immediately after "France" and "Poland" is an *omission comma*—it takes the place of the missing words "is famous." See also *Editing II – Punctuation Highlights*. Choice B provides an example of a run-on sentence. There must be an "and" preceding the word "Poland." As it stands, it is three sentences joined together by commas.

Choice C contains two sentence fragments: "France for its chefs and philosophers" and "Poland for its mathematicians and logicians." These phrases cannot stand on their own as complete sentences. Choice D improperly uses the pronoun "their," when what is called for is the pronoun "its." Moreover, we would need to have commas after both the words "France" and "Poland" in order to validate this choice; alternatively, we could omit commas after France and Poland. Words can be omitted within a sentence if they're a readily understood in context.

Choice E changes the meaning of the original sentence (that's a no-no). There's little doubt that France and Poland have composers, musicians, chefs, philosophers, mathematicians, and logicians, but the focus is on what each country is specifically famous for.

In summary, there are four possible correct answers.

Correct   Italy is famous for its composers and musicians, France is famous for its chefs and philosophers, and Poland is famous for its mathematicians and logicians.

Note that the above version repeats three times the words "is famous."

Correct   Italy is famous for its composers and musicians, France, for its chefs and philosophers, and Poland, for its mathematicians and logicians.

The above is the correct rendition per choice A. The comma after "France" and "Poland" is effectively taking the place of the words "is famous."

Correct   Italy is famous for its composers and musicians, France for its chefs and philosophers, and Poland for its mathematicians and logicians.

The above version is likely the most subtle. The rules of ellipsis allow us to omit words that are readily understood within the context of any sentence. The words "is famous" are readily understood. This version is almost identical to choice B, except that it correctly inserts the word "and" (a conjunction) before Poland.

Correct   Italy is famous for its composers and musicians; France, for its chefs and philosophers; Poland, for its mathematicians and logicians.

The above version uses semicolons along with commas. Note that the final "and" (conjunction) before Poland is optional. Unlike choice D, this choice correctly inserts a comma after "France" and "Poland" and replaces the pronoun "their" with "its."

16. Sweater

Choice E
Classification: Comparisons
Skill Rating: Easy
Snapshot: This problem is included to highlight the handling of the comparative and superlative adjective forms.

The words "neither one" indicate that we are dealing with two sweaters. When comparing two things, we use the comparative form of the adjective, not the superlative. Thus, the correct choice is "better," not "best," and "smaller," not "smallest." "Better" and "smaller" (comparatives) are used when comparing exactly two things; "best" and "smallest" (superlatives) are used when comparing three or more things.

NOTE ೞ When two things are being compared, the *comparative* form of the adjective (or adverb) is used. The comparative is formed in one of two ways: (1) adding "er" to the adjective (for adjectives containing one syllable), or (2) placing "more" before the adjective (especially for adjectives with more than two syllables). Use one of the above methods, but never both: "Jeremy is wiser (or *more wise*) than we know," but never "Jeremy is more wiser than we know."

When three or more things are being compared, the *superlative* form of the adjective (or adverb) is used. The superlative is formed in one of two ways: (1) adding "est" to the adjective (for adjectives containing one syllable), or (2) placing "most" before the adjective (especially for adjectives with more than two syllables). Use one of the above methods, but never both: "He is the cleverest (or *most clever*) of my friends," but never "He is the most cleverest of my friends."

Some modifiers require internal changes in the words themselves. A few of these irregular comparisons are presented in the following chart:

Positive	Comparative	Superlative
good	better	best
well	better	best
bad	worse	worst
far	farther, further	farthest, furthest
late	later, latter	latest, last
little	less, lesser	least
many, much	more	most

17. Sir Isaac Newton

Choice E
Classification: Comparisons
Skill Rating: Easy
Snapshot: This solution to this problem pivots on the use of the demonstrative pronoun "those."

The words "those" and "other" must show up in the correct answer. Without the word "other," choices A, C, and D illogically compare Sir Isaac Newton to all scientists, living or dead, even though Sir Isaac Newton is one of those scientists. Without the word "those," choices A and B illogically compare "the accomplishments of Sir Isaac Newton" to "other scientists." Obviously, we must compare "the accomplishments of Sir Isaac Newton" to "the accomplishments of other scientists." In choices C, D, and E, the word "those" exists to substitute for the phrase "the accomplishments."

PUTTING IT ALL TOGETHER

18. Soya

Choice C
Classification: Comparisons
Skill Rating: Medium
Snapshot: This problem highlights the use of the demonstrative pronoun "that."

Here, we must correctly compare "the protein in meat" to "the protein in soybeans." The demonstrative pronoun "that" is very important because it substitutes for the words "the protein." Choice C creates a sentence which effectively reads: "In addition to having more protein than meat does, the protein in soybeans is higher in quality than *the protein* in meat."

Choices A and B are out because the word "meat" must come after the opening phrase "in addition to having more protein than meat does." Choice D correctly employs "soybeans" but incorrectly uses "it" to make a comparison. The word "it" cannot stand for "the protein." Choice E incorrectly compares soybean protein to meat.

19. Angel

Choice C
Classification: Comparisons
Skill Rating: Medium
Snapshot: This problem is included to highlight proper comparisons involving "like" versus "as."

The basic difference between "like" and "as" is that "like" is used for phrases, and "as" is used for clauses. A phrase is a group of words that does not contain a verb; a clause is a group of words that does contain a verb. Choices D and E ungrammatically employ "as" in phrases, in addition to being awkwardly constructed.

There are three potentially correct versions:

THE LITTLE GOLD GRAMMAR BOOK

1) She sings <u>like an angel</u>.
"Like an angel" is a phrase (there is no verb), so "like" is the correct choice.

2) She sings <u>as an angel sings</u>.
"As an angel sings" is a clause (contains the verb "sings"), so "as" is the correct choice.

3) She sings <u>as an angel does</u>.
"As an angel does" is a clause (contains the verb "does"), so "as" is the correct choice.

NOTE ○ Advertising is an arena where violations in English grammar may be turned to advantage. The American cigarette company Winston once adopted the infectious advertising slogan: "Winston tastes good like a cigarette should." The ungrammatical and somehow proactive use of "like" instead of "as" created a minor sensation, helping to propel the brand to the top of the domestic cigarette market. A more recent advertising campaign by DHL in Asia also contains a grammatical violation: "No one knows Asia like we do." The correct version should read: "No one knows Asia as we do."

20. Perceptions

Choice D
Classification: Comparisons
Skill Rating: Medium
Snapshot: This problem is included to highlight the comparative idiom "as...do"/"as...does."

The problem pivots on the "like/as" distinction. If the underlined portion modifies "right-brained individuals," then "like" would be appropriate; however, if it parallels the clause "right-brained individuals do not employ," then "as" is appropriate. To modify "right-brained individuals," the underline should be next to the word, so choices A and B are not correct. Also, choice A states that "right-brained individuals" and "left-brained individuals" are similar, whereas the rest of the sentence contrasts them. Choices C, D, and E use the correct connector, "as," but choice E, like choice B, uses the singular "adult," and choices C and E employ phrases as opposed to clauses. Choice E provides the proper comparative clause.

Choice B is contextually sound but structurally awkward. Either of the following would be better:

Correct	Unlike left-brained individuals, right-brained individuals often do not employ their attention or perceptions systematically, and they may not notice and remember the same level of detail <u>as</u> their left-brained counterparts <u>do</u>.
Correct	Right-brained individuals often do not employ their attention or perceptions systematically, and, unlike left-brained individuals, right-brain individuals may not notice and remember the same level of detail <u>as</u> their left-brained counterparts <u>do</u>.

21. Geography

Choice A
Classification: Comparisons
Skill Rating: Medium
Snapshot: This problem is included to highlight the correct use of the "more...than" idiom, used in comparing two things.

Make an initial note that we should ideally be comparing American high school *students* with Canadian high school *students* (plural with plural) because the non-underlined part of the sentence contains the words "counterparts." Be suspicious of any of the answer choices which begin with "the American high school student." Verify also that in all cases verbs are correct. "Do" is a plural verb that matches the plural phrase "Canadian counterparts"; "does" is a singular verb that would be used to match the singular phrase "Canadian counterpart."

The last piece of the puzzle is to eliminate the non-standard comparative constructions, namely "more...compared to" as well as "more...compared with." The correct idiom is "more...than" or "less...than." Thus, choices B, C, and E cannot be correct. See chapter 3 for a list of the *200 Common Grammatical Idioms*.

22. Assemblée Nationale

Choice D
Classification: Comparisons
Skill Rating: Medium
Snapshot: This problem is included to highlight the comparative idiom "Just as...so (too)." Note that the brackets indicate the optional use of the word "too."

In choices A and B, the use of "as" is incorrect. "As" functions as a subordinating conjunction, and this means that the reader expects a logical connection between the fact that Britain has a Parliament and France has the Assemblée Nationale. Try substituting the subordinating conjunction "because" in either choices A or B and the illogical relationship becomes more apparent. "Because Parliament is the legislative government body of Great Britain, the Assemblée Nationale is the legislative government body of France."

The "just as...so (too)" comparative idiom (choice D) can be used to express this type of meaning. "Just as something, so something else." Choice D provides a standard comparison: The Parliament of Great Britain is being compared to the Assemblée Nationale of France. In choice E, the comparison is awkward because we end up comparing the Government of Britain's Parliament with the Assemblée Nationale.

Choice C is not only awkward, but "just like" also is not correct; it is a redundancy where "like" would otherwise do the trick. "Like" is used for phrases, whereas "as" is used for clauses. Clearly we are dealing with a clause.

NOTE ✑ Savor this classic example:

Correct   <u>Just as</u> birds have wings, <u>so too</u> do fish have fins.

Incorrect   As birds have wings, fish have fins.

Incorrect   As birds have wings, fish, therefore, have fins.

Substituting "because" for "as" above, we can quickly see an illogical relationship. There is no connection between a bird's having wings and a fish's having fins.

Incorrect   Just like birds that have wings, fish have fins.

We can't use "just like" because "like" is not used with clauses; "that have wings" is a clause. Moreover, "just like" is considered an unnecessary redundancy of "like." So although "just as" is grammatically sound, we really shouldn't use "just like" in formal writing.

23. Bear

Choice C
Classification: Comparisons
Skill Rating: Difficult
Snapshot: When making comparisons, the most basic rule is to make sure to compare like things. That is, compare apples with apples and oranges with oranges. This is particularly true when distinguishing between the characteristics of one thing to the characteristics of something else. In such cases, we must compare thing to thing, and characteristic to characteristic.

Here we want to compare "bears" with "bears" or "diets of bears" with "diets of bears." Choice A, the original, compares animals with diets by erroneously comparing the "Alaskan brown bear and most other members" of the bear family to the "diet" of the grizzly bear. Choice B is structurally sound ("those" is a demonstrative pronoun that takes the place of "the diets") but unidiomatically refers to the "diets" of the grizzly bear. Idiomatic speech would require the use of "diet" to refer to a single bear species and "diets" to refer to more than one species of bear. Choice D uses the repetitious "Just like" (when "Like" alone is sufficient), as well as the unidiomatic "diets." Choice E commits the original error in reverse. Now "diets" of the Alaskan brown bear and most other members of the bear family are being compared directly to "grizzly bears."

All of the following provide potentially correct answers:

1) Like the Alaskan brown bear and most other members of the bear family, the grizzly bear has a diet consisting of both meat and vegetation.

2) Like the Alaskan brown bear and most other members of the bear family, grizzly bears have a diet consisting of both meat and vegetation.

3) Like the diets of the Alaskan brown bear and most other members of the bear family, the diet of the grizzly bear consists of both meat and vegetation.

4) Like the diets of the Alaskan brown bear and most other members of the bear family, the diet of grizzly bears consists of both meat and vegetation.

24. Smarts

Choice E
Classification: Comparisons
Skill Rating: Difficult
Snapshot: This problem is included as an "oddball" to demonstrate that we do not always compare a singular item with singular item or plural item with plural item (e.g., Miller Analogies Test versus

IQ tests). In context, a situation may necessitate comparing a singular item with a plural item or vice versa. Here the "apples to apples, oranges to oranges" comparison involves comparing one type of test to another type of test while comparing the formats of one such test to the formats of the other types of tests.

Choices A and B erroneously compare "the Miller Analogies Test" with "the formats..." We want to compare "one exam" to "another exam," or "the format of one exam" to the "format of another exam," or "the formats of some exams" to the "formats of other exams." Although choice C looks like the winning answer, upon closer examination we realize that a single format cannot itself vary considerably in terms of content and length. Choice D correctly employs "formats," but now the problem reverses itself: A single IQ test does not have "formats." Choice E correctly combines "IQ tests" in the plural with "formats" in the plural.

Here's a follow-up example in mirror image to the problem at hand:

Incorrect	Unlike Canadian football, which is played on a standardized field, American baseball is played on a <u>field</u> which varies considerably in shape and size.
Correct	Unlike Canadian football, which is played on a standardized field, American baseball is played on <u>fields</u> which vary considerably in shape and size.

25. Golden Years

Choice E
Classification: Verb Tenses
Skill Rating: Easy
Snapshot: This problem is included to illustrate the difference between the present perfect tense and the past perfect tense. The correct answer employs the present perfect tense.

Only choice E uses the correct tense (present perfect), observes parallelism, and is idiomatic. Because the sentence describes a situation that continues into the present, choices A and B are incorrect in using the past perfect tense ("had elected"). In choice E, the noun forms "to retire" (infinitive) and "face" are more closely parallel than are the noun forms "retirement" and "facing." Note also that the dual expressions "x rather than y" and "x instead of y" are, according to Standard Written English, equivalent.

26. Politics

Choice D
Classification: Verb Tenses
Skill Rating: Medium
Snapshot: This problem is included to highlight the past perfect tense and the precise use of the auxiliary "had" in forming this tense.

The original sentence contains two critical past tense verbs: "disapproved" and "conceded." It also contains the time word "later," as in "later conceded," which serves to further clarify the sequence of past events. This problem highlights an important characteristic of the past perfect tense, namely that "had" is used before the first of two past events. In this example, Senator John McCain "disapproved" before he "conceded." Thus, the auxiliary "had" must be placed before the first (not the second) of the two past events: "had disapproved...later conceded."

Choice A erroneously proposes a reversal in sequence ("disapproved...had later conceded"), while choice E doubles the use of "had" to create a verbal muddle ("had disapproved...had later conceded"). Both of these choices result in illogical alternatives. Choices B and C incorrectly employ the present perfect tense ("has") when the past perfect tense ("had") is what is called for.

NOTE ○ Another correct answer would have included the following:

"Although he disapproved of the political platform set forth by Senator Barack Obama during the 2008 U.S. presidential primaries, Senator John McCain later conceded..."

This option is also correct, although it doesn't use the past perfect tense. It instead uses two past tense verbs, namely "disapproved" and "conceded," and the temporal word "later." Because the sequence of tense is clear, the use of the auxiliary "had" is considered optional. Refer to the explanation given for Q75 in *Answers to the 100 Question Quiz.*

27. Trend

Choice A
Classification: Verb Tenses
Skill Rating: Medium
Snapshot: This problem is included to illustrate the difference between the simple past tense versus the past perfect tense and the present perfect tense. The correct answer uses the simple past tense.

Here, the simple past tense is all that is needed to refer clearly to the time frame in the past (1980–1990). In choice B, the present perfect tense "has continued" is inconsistent with the timing of an event that took place in the distant past. In choice C, the past perfect tense "had continued" is not required because we are not making a distinction between the sequence of two past tense events.

In choices D and E, the focus switches from a rise in the "percentage of people" to a rise in the "abrupt increase." This shift in meaning is unwarranted and incorrect. The pronouns "it" (choice D) and "which" (choice E) are ambiguous and could refer to either the "percentage of people" or an "abrupt increase." Moreover, choices D and E employ the passive constructions "there had been" and "there was"; these are considered weak sentence constructions and are best avoided.

28. Fire

Choice B
Classification: Verb Tenses
Skill Rating: Medium
Snapshot: This problem is included to highlight the difference between the past perfect tense and the simple past tense and/or the present perfect tense. The correct answer employs the past perfect tense. This problem also addresses the passive verb construction "had been"/"have been."

The solution to this problem is conceptually similar to that of the preceding problem. The auxiliary "had" must be used in conjunction with the first of two past tense events. In short, only choice B uses the verb tenses correctly to indicate that houses were built or heavily damaged prior to their being destroyed by fire. Choices A, C, and E illogically state that some houses were both destroyed <u>and</u> heavily damaged; "or" is needed to indicate that each of the houses suffered either one fate or the other. In using only the simple past tense (i.e., the verb tense "were"), choice A fails to indicate that the houses were built before the fires occurred. Choices D and E erroneously employ the present perfect tense, saying in effect that the houses "have been constructed" after they were destroyed or heavily damaged last year.

29. B-School

Choice E
Classification: Verb Tenses
Skill Rating: Medium
Snapshot: This problem is included to illustrate the difference between the simple future tense and the present perfect tense (both simple and progressive verb forms). The correct answer uses the simple future tense.

Since all answer choices contain the words "in the coming years," we definitely know we are dealing with the future, and choice E complements our search for a simple future tense. In choices A and B, the tense "have been becoming" (present perfect progressive tense in the passive voice) doesn't work. In choices C and D, the present perfect tense is also out. The present perfect tense is useful only for events that began in the past and touch the present. Here we need a tense that takes us into the future.

30. Summer in Europe

Choice C
Classification: Verb Tenses
Skill Rating: Medium
Snapshot: This problem demonstrates the correct use of the future perfect tense.

This problem requires the use of the future perfect tense. Choices A and B, by employing the construction "have reached," offer incorrect versions based on the present perfect tense. Choices D and E create erroneous alternatives by commingling past tense constructions with those in the future tense. Choice D presents an incorrect version which doubles up the present perfect tense "have reached" with the future perfect tense "will have backpacked."

Choice E mixes the simple past tense "reached" with the future perfect tense (in the progressive form). For the record, an equally correct answer would have been: "By the time we reached France, we had been backpacking for 12 weeks." This would represent the correct use of the past perfect tense. Of course, the original sentence clearly indicates that the travelers are looking into the future—they have not yet arrived in France.

The future perfect tense and the past perfect tense are very much opposite in terms of time frame but structurally similar.

Past perfect tense: By the time something happened (second event), something else had already happened (first event).

Future perfect tense: By the time something happens (second event), something else will have already happened (first event).

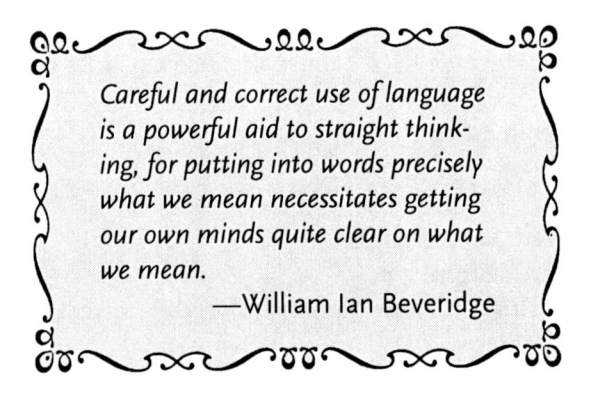

> *Careful and correct use of language is a powerful aid to straight thinking, for putting into words precisely what we mean necessitates getting our own minds quite clear on what we mean.*
> —William Ian Beveridge

Editing I – Tune-up

Rarely does our educational system, or writing skills courses in the particular, include a segment on editing. Editing is to writing what an oil change and tune-up, washing, waxing, vacuuming, and chamoising are to automobile care. Not only does it influence how we feel about the final product, but it directly impacts how others perceive our work. Editing is its own skill set. It's an integral writing component that demands a separate, dedicated review.

a vs. an. Use "a" before a word in which the first letter of that word is a consonant or has the sound of a consonant. Note that some vowels have the sound of a consonant when pronounced as individual letters.

Example	a fortune (the letter "f" is a consonant)
Example	a B.S. degree (the letter "B" is a consonant)
Example	a u-turn (the letter "u" is pronounced "yoo")

Use "an" before a word in which the first letter of that word is a vowel or has the sound of a vowel. Note that some consonants sound like vowels when pronounced as individual letters.

Example	an ox (the letter "o" is a vowel)
Example	an M.S. degree (the letter "M" is pronounced "em")
Example	an honor (the letter "h" is silent so "an" is matched with the letter "o"—a vowel)

Abbreviations (Latin). The abbreviations "e.g." (meaning "for example") and "i.e." (meaning "that is") are constructed with two periods, one after each of the two letters, with a comma always following the second period. The forms "eg." or "ie." are not correct.

Below are three ways to present information using "for example."

Correct	A number of visually vibrant colors (e.g., orange, pink, and purple) are not colors that would normally be used to paint the walls of your home.

Exhibit A Latin Abbreviations and Their Meaning

Abbreviations	Meaning
c.	approximately
cf.	compare
e.g.	for example
etc.	and so forth
et al.	and others
ibid.	in the same place
i.e.	in other words; that is
op. cit.	in the work cited
sc.	which means
sic.	in these exact words
s.v.	under the word or entry
v.	consult
viz.	namely

Correct   A number of visually vibrant colors, e.g., orange, pink, and purple, are not colors that would normally be used to paint the walls of your home.

Correct   A number of visually vibrant colors, for example, orange, pink, and purple, are not colors that would normally be used to paint the walls of your home.

Below are three ways to present information using "that is."

Correct   The world's two most populous continents (i.e., Asia and Africa) account for 75 percent of the world's population.

Correct   The world's two most populous continents, i.e., Asia and Africa, account for 75 percent of the world's population.

Correct   The world's two most populous continents, that is, Asia and Africa, account for 75 percent of the world's population.

The Latin abbreviations as listed in *Exhibit A* should be used with caution. Their use depends on whether the intended audience is likely to be familiar with their meaning. This compilation is not so much an endorsement for their use as it is a convenient list in case readers find them in various works.

NOTE ଔ The abbreviation "etc." stands for "et cetera" and translates as "and so forth." Never write "and etc." because "and" is redundant and otherwise reads "and and so forth." See *Appendix II – American English vs. British English* for additional coverage of abbreviations.

Apostrophes for omitted letters. When apostrophes are used to represent omitted letters, they are always "nines" not "sixes." This means they curl backwards not forwards.

Example   rock 'n' roll (not rock 'n' roll)

Example   jivin' (not jivin')

Example   'tis (not 'tis)

Brevity. As a general rule, less is more. Consider options that express the same ideas in fewer words without changing the meaning of a sentence.

Less effective	A movie director's <u>skill</u>, <u>training</u>, and <u>technical ability</u> cannot make up for a poor script.
More effective	A movie director's <u>skill</u> cannot make up for a poor script.

Often you can cut "of" or "of the."

Original	employees of the company
Better	company employees

Don't use "due to the fact that" or "owing to the fact that." Use "because" or "since."

Original	<u>Owing to the fact that</u> questionnaires are incomplete, it is difficult to draw definitive conclusions.
Better	<u>Because</u> questionnaires are incomplete, it is difficult to draw definitive conclusions.
Original	We want to hire the second candidate <u>due to the fact that</u> he is humorous and has many good ideas.
Better	We want to hire the second candidate <u>since</u> he is humorous and has many good ideas.

Bullets vs. hyphens or asterisks. Bullets are most commonly used with résumés and flyers, but they are also welcomed companions in nonfiction, especially when used with lists or tables. It is not considered good practice in formal writing to use hyphens (-) or asterisks (*) in place of bullets. Standard protocol requires use of round bullets, square bullets, or perhaps webdings, wingdings, or dingbats—these "dings" represent ornamental bullets or tiny graphical characters.

THE LITTLE GOLD GRAMMAR BOOK

Examples include: ●, •, ○, ■, ▪, ◆, ♦, ◊, ▶, ▼, ❑, ➪, ➢

Bulleted lists. Displaying information vertically is often done with bulleted lists or numbered lists. The following provides a succinct summary on how to punctuate bulleted (or numbered) information. There are five basic scenarios. The topic chosen is one dear to us all—books!

Scenario 1A

Four words that describe why books are cool:

- durability
- accessibility
- portability
- affordability

Treatment: Here, a complete sentence introduces a bulleted list and it is appropriately followed by a colon. No periods follow any line of bulleted information because none is a complete sentence. Capitalization of these words is optional; however, each of these words should begin with a capital letter or each word should be placed in lowercase (no capital letters).

Scenario 1B

Four reasons why printed books are great:

1. Durability
2. Accessibility
3. Portability
4. Affordability

Treatment: Numbered lists are punctuated identically to bulleted lists. The one exception is that the first word that follows each number must be capitalized. Note that numbered lists are not recommended unless there is a reason for their use, including the need to impose order or hierarchy.

Scenario 2

Why will printed books never become obsolete?

- They're durable.
- They're accessible.
- They're portable.
- They're affordable.

Treatment: The complete sentence that introduces the list above is appropriately followed, in context, by a question mark, not a colon. Since each line of bulleted information is a complete sentence, each begins with a capital letter and ends with a period.

Scenario 3

People love printed books because they're

- durable
- accessible
- portable
- affordable

Treatment: No colon is used after the word "they're" because it does not introduce a complete sentence. No periods follow any of the bulleted information because none is a complete sentence. Capitalization of the first word following each bullet is optional.

Scenario 4

Printed books are here to stay because

- They're highly durable.
- They're easily accessible.
- They're wonderfully portable.
- They're eminently affordable.

Treatment: The bulleted information above is not introduced by a complete sentence and no colon is used after the word "because." The beginning word of each bullet point is capitalized and a period ends each complete sentence.

Scenario 5

People love printed books because they're

- highly durable;
- easily accessible;
- wonderfully portable;
- eminently affordable.

Treatment: Because the bulleted information reads as a single sentence, it is possible to use a semicolon to separate bulleted information and use a period on the last line. However, if bulleted information is short, as in the examples above, it may be better to enumerate the list in run-in text rather than displaying information vertically. For example:

People love printed books because they're (1) highly durable, (2) easily accessible, (3) wonderfully portable, and (4) eminently affordable.

A note on using periods: Periods are often used with bulleted information, as is the case with résumés or slide presentations. Inconsistency arises when periods are used arbitrarily, appearing at the end of one bulleted point but not another. Again, the "hard" rule is to put a period at the end of any bulleted information that forms a complete sentence and omit any period at the end of any bulleted information that does not form a complete sentence. However, an arguably more practical rule of thumb with respect to résumés or slide presentations would be to omit a period after any short bulleted information (say six or fewer words) and include periods after any bulleted information that extends more than a line in length (regardless of whether or not it forms a complete sentence). The rationale for erring on the side of including periods is that a period helps create closure for the eye, thereby enhancing readability.

Colon. A colon (a punctuation mark that consists of two vertical dots) is not used after the words *namely, for example, for instance,* or *such as*. When introducing a list or series of items, a colon is not used after forms of the verb "to be" (i.e., *is, are, am, was, were, have been, had been, being*) or after "short" prepositions (e.g., *at, by, in, of, on, to, up, for, off, out, with*).

Incorrect	We sampled several popular cheeses, namely: Gruyere, Brie, Camembert, Roquefort, and Stilton.
	(Remove the colon placed after the word "namely.")
Incorrect	My favorite video game publishers are: Nintendo, Activision, and Ubisoft.
	(Remove the colon placed after the verb "are.")
Incorrect	Graphic designers should be proficient at: Photoshop, Illustrator, InDesign, and Adobe Acrobat.
	(Remove the colon placed after the preposition "at.")

However, if what follows a colon is not a list or series of items, the writer is free to use the colon after any word that he or she deems fit.

Correct	The point is: People who live in glass houses shouldn't throw stones.

(A colon follows the verb "is.")

Correct	Warren Buffett went on: "Only four things really count when making an investment—a business you understand, favorable long-term economics, able and trustworthy management, and a sensible price tag. That's investment. Everything else is speculation."

(A colon follows the preposition "on.")

Compound adjectives. Compound adjectives (also called compound modifiers) occur when two (or more) words act as a unit to modify a single noun. As illustrated in *Exhibit B,* use a hyphen to join the compound adjective when it comes before the noun it modifies, but not when it comes after the noun.

Note that in situations where compound adjectives are formed using multiple words and/or words that are already hyphenated, it is common practice to use an en dash (–) to separate them. See entry under *Dashes*.

Example	Los Angeles–Buenos Aires

Example	quasi-public–quasi-private medicare bill

Sometimes compound adjectives consist of a string of "manufactured" words.

Example	a fly-by-the-seat-of-your-pants entrepreneur

Exhibit B   Compound Adjectives

Compound adjectives (hyphenated)	Non-compound adjectives (not hyphenated)
Experience teaches a person to use a step-by-step approach when solving problems. "Step by step" comes before the noun "approach," so it is hyphenated.	Experience teaches a person to approach solving problems step by step. "Step by step" comes after the noun "approach," so it is not hyphenated.
Send a follow-up e-mail.	Send an e-mail to follow up.
Write an in-depth report.	The report discussed the topic in depth.
An informative and up-to-date newsletter.	A newsletter that is informative and up to date.
A well-known person.	A person who is well known.
A well-intentioned act.	An act that is well intentioned.
Five brand-new bikes.	Five bikes that are brand new.
The ten-year-old girl.	The girl who is ten years old.
A thirty-five-year-old CEO.	A CEO who is thirty-five years old.

Example   a tell-it-like-it-is kind of spokesperson

There are four potentially confusing situations where compound adjectives are either not formed or not hyphenated. The first occurs where a noun is being modified by an adjective and an adjective is being modified by an adverb.

Example   very big poster

In the previous example, "big" functions as an adjective describing the noun "poster," and "very" functions as an adverb describing the adjective "big."

The second situation occurs when adjectives describe a compound noun: that is, two words that function as a single noun.

Example   cold roast beef

Here the word "cold" functions as an adjective to describe the compound noun "roast beef." We would not write "cold-roast beef" because "cold-roast" does not jointly modify "beef."

Example   little used book

Here the word "little" functions as an adjective to describe the compound noun "used book." The meaning here is that the book is not new and also little. However, it would also be correct to write "little-used book" if our intended meaning was that the book was not often referred to.

A third situation occurs when a compound noun describes another noun.

Example   high school student

Example   cost accounting issues

"High school" is considered a compound noun that describes "student." Compound nouns are not hyphenated. "Cost accounting," which describes "issues," is also a non-hyphenated compound noun.

A fourth situation occurs when compounds are formed with adverbs ending in "ly." Adverbs ending in "ly" are not hyphenated, even when functioning as compound modifiers.

Example   a highly motivated employee

Example   a newly published magazine

Example   a publicly traded company

Example   a frequently made error

NOTE ☙ "Family-owned" and "family-run" are hyphenated (when functioning as compound adjectives) because "family," although ending in "ly," is not an adverb.

Dashes. Note first the difference between a hyphen and a dash. A dash is longer than a hyphen (-) and should not be used when what is needed is a dash. There are two types of dashes. The first is called "em dash" ("—"), which is the longer of the two dashes. The second is called "en dash" ("–"), which is the shorter of the two dashes. The en dash (–) is most popular in everyday writing, while the em dash (—) is the standard convention for formal published documents. Incidentally, the en dash is so-called because it is the width of the capital letter "N"; the em dash is so-called because it is the width of the capital letter "M." These two types of dashes can be found in Microsoft Word® under the pull-down menu Insert, Symbols, Special Characters.

Three common conventions arise relating to the use of the dash: (1) an en dash (" – ") with spaces on both sides of the dash, (2) an em dash ("—") with no spaces on either side of the dash, and (3) an em dash (" — ") with spaces on both sides of the dash. The first two conventions are the most popular for written (non-published) documents. The third option is popular on websites.

Example   To search for wealth or wisdom – that's a classic dilemma.

                (Spaces on both sides of the en dash.)

Example   To search for wealth or wisdom—that's a classic dilemma.

(No space on either side of the em dash.)

Example   To search for wealth or wisdom — that's a classic dilemma.

(Space on both sides of the em dash.)

Hyphen. Use a hyphen with compound numbers between twenty-one through ninety-nine and with fractions.

Example   Sixty-five students constitute a majority.

Example   A two-thirds vote is necessary to pass.

In general, use a hyphen to separate component parts of a word in order to avoid confusion with other words especially in the case of a double vowel.

Example   Our goal must be to <u>re-establish</u> dialogue, then to <u>re-evaluate</u> our mission.

Example   Samantha's hobby business is turning <u>shell-like</u> ornaments into jewelry.

Use a hyphen to separate a series of words having a common base that is not repeated.

Example   small- to medium-sized companies

This of course is the shortened version of "small-sized to medium-sized companies."

Example   short-, mid-, and long-term goals

This is the shortened version of "short-term, mid-term, and long-term goals."

In general, use hyphens with the prefixes *ex-* and *self-* and in forming compound words with *vice-* and *elect-*.

Example   Our current vice-chancellor, an ex-commander, is a self-made man.

NOTE ☙ "Vice president" (American English) is not hyphenated, but "vice-presidential duties" is.

Nominalizations. A guiding rule of style is that we should prefer verbs (and adjectives) to nouns. Verbs are considered more powerful than nouns. In other words, a general rule in grammar is that we shouldn't change verbs (or adjectives) into nouns. The technical name for this no-no is "nominalization"; we shouldn't nominalize.

Avoid changing verbs into nouns:

More effective	reduce costs
Less effective	reduction of costs
More effective	develop a five-year plan
Less effective	development of a five-year plan
More effective	rely on the data
Less effective	reliability of the data

In the above three examples, the more effective versions represent verbs, not nouns. So "reduction of costs" is best written "reduce costs," "development of a five-year plan" is best written "develop a five-year plan," and "reliability of the data" is best written "rely on the data."

Avoid changing adjectives into nouns:

More effective	precise instruments
Less effective	precision of the instruments

More effective	<u>creative</u> individuals
Less effective	<u>creativity</u> of individuals
More effective	<u>reasonable</u> working hours
Less effective	<u>reasonableness</u> of the working hours

In the latter three examples above, the more effective versions represent adjectives, not nouns. So "precision of instruments" is best written "precise instruments," "creativity of individuals" is best written "creative individuals," and "reasonableness of the working hours" is best written "reasonable working hours."

Numbers. The numbers one through one hundred, as well as any number beginning a sentence, are spelled out. Numbers above 100 are written as numerals (e.g., 101).

Original	Our professor has lived in 3 countries and speaks 4 languages.
Correct	Our professor has lived in <u>three</u> countries and speaks <u>four</u> languages.

Page numbering. A time-honored convention in publishing is that odd-numbered pages are "right-hand" pages and all pages are counted, whether or not a page number is printed on a page. In the case of a nonfiction book, this means that page 1 is the title page (no page number is printed), page 2 is the copyright page (no page is number printed), page 3 is the table of contents (the page number may or may not be printed), page 5 is the introduction (the page number is printed), and so forth.

In the case of a business report, page 1 is the title page (no page number is printed), page 3 is often the executive summary (the page number may or may not be printed), page 5 is the table of contents (the page number may or may not be printed), page 7 is the introduction (the page number is printed), and so forth.

Obviously the exact types of information included in a book, report, or academic research paper will vary, but three page-formatting conventions will always be adhered to. First, odd-numbered pages will always be right-hand or front pages and even-numbered pages will always be left-hand or backside pages. Second, all pages will count toward the total number of pages. Third, all new sections begin as odd-numbered, right-hand pages (with few exceptions). This means that if one section ends on an odd-numbered page, then the next page will be "skipped" so that the next section can begin on an odd-numbered page. The page that was skipped (an even numbered, left-hand page) remains blank, and although no page number is printed on it, it tallies in the page count as would an actual, fully printed page. Following these three conventions helps ensure that long documents look professionally laid out.

Paragraph styles. Two basic formats may be followed when laying out a written document: "block-paragraph" format and "indented-paragraph" format. The block-paragraph format typifies the layout of the modern business letter. Each paragraph is followed by a single line space (one blank line). Paragraphs are blocked, meaning that every line aligns with the left-hand margin with no indentation. Often, paragraphs are fully justified, which means there are no "ragged edges" on the right-hand side of any paragraph.

The indented-paragraph format is the layout followed in a novel. The first line of each paragraph is indented and there is no line space used between paragraphs within a given chapter. Note, however, that the first line of opening paragraphs, those that begin a new chapter, are not indented (they are left justified).

Whereas the indented-paragraph format (with indented paragraphs) usually has the effect of making writing look more personable—more like a story—the block-paragraph format (typically with fully justified paragraphs) lends a more formal appearance.

Passive voice vs. active voice. As a general rule of style, write in the active voice, not in the passive voice (all things being equal).

Less effective	Sally <u>was</u> loved by Harry.
More effective	Harry loved Sally.
Less effective	In pre-modern times, medical surgery <u>was</u> often performed by inexperienced and ill-equipped practitioners.
More effective	In pre-modern times, inexperienced and ill-equipped practitioners often performed medical surgery.

In a normal subject-verb-object sentence, the doer of the action appears at the front of the sentence while the receiver of the action appears at the back of the sentence. Passive sentences are less direct because they reverse the normal subject-verb-object sentence order; the receiver of the action becomes the subject of the sentence and the doer of the action becomes the object of the sentence. Passive sentences may also fail to mention the doer of the action.

Less effective	Errors <u>were</u> found in the report.
More effective	The report contained errors.
or	The <u>reviewer</u> found errors in the report.
Less effective	Red Cross volunteers should <u>be</u> generously praised for their efforts.
More effective	<u>Citizens</u> should generously praise Red Cross volunteers for their efforts.
or	<u>We</u> should generously praise Red Cross volunteers for their efforts.

How can we recognize a passive sentence? Here's a quick list of six words that signal a passive sentence: *be, by, was, were, been,* and *being*. For the record, "by" is a preposition, not a verb form, but it frequently appears in sentences that are passive.

Possessives. Confusion can arise regarding how to create possessives with respect to nouns. There are four basic situations. These involve (1) creating possessives with respect to single nouns not ending in "s"; (2) creating possessives with respect to single nouns ending in "s"; (3) creating possessives for plural nouns not ending in "s"; and (4) creating possessives for plurals ending in "s."

For <u>single nouns</u> not ending in the letter "s," we simply add an apostrophe and the letter "s" (i.e., 's).

| Example | Jeff's bike |
| Example | The child's baseball glove |

For <u>single nouns</u> ending in the letter "s," we have a choice of either adding an apostrophe and the letter "s" (i.e., 's) or simply an apostrophe.

| Example | Professor Russ's lecture |
| Example | Professor Russ' lecture |

For <u>plural nouns</u> not ending in the letter "s," we simply add an apostrophe and the letter "s" (i.e., 's).

| Example | men's shoes |
| Example | children's department |

For <u>plural nouns</u> ending in the letter "s," we simply add an apostrophe. Note that most plural nouns do end in the letter "s."

| Example | ladies' hats |
| Example | The boys' baseball bats |

In this latter example, "boys' baseball bats" indicates that a number of boys have a number of (different) baseball bats. If we were to write "boys' baseball bat," it would indicate that a number of boys all own or share the same baseball bat. If we wrote "the boy's baseball bat," only one boy would own the baseball bat. In writing "the boy's baseball bats," we state that one boy possesses several baseball bats.

Print out to edit. Do not perform final edits on screen. Print documents out and edit from a hard copy.

Qualifiers. Whenever possible, clean out qualifiers, including: *a bit, a little, fairly, highly, just, kind of, most, mostly, pretty, quite, rather, really, slightly, so, still, somewhat, sort of, very,* and *truly.*

Original	Our salespeople are just not authorized to give discounts.
Better	Our salespeople are not authorized to give discounts.
Original	That's quite a big improvement.
Better	That's a big improvement.
Original	Working in Reykjavik was a most unique experience.
Better	Working in Reykjavik was a unique experience.

NOTE ଔ Unique means "one of a kind." Something cannot be somewhat unique, rather unique, quite unique, very unique, or most unique, but it can be rare, odd, or unusual.

Quotations. The following four patterns are most commonly encountered when dealing with quotations.

Example	My grandmother said, "An old picture is like a precious coin."
Example	"An old picture is like a precious coin," my grandmother said.
	(A comma is generally used to separate the quote from regular text.)
Example	"An old picture," my grandmother said, "is like a precious coin."
	(Above is what is known as an interrupted or split quote. The lower case "i" in the word "is" indicates that the quote is still continuing.)
Example	"They're like precious coins," my grandmother said. "Cherish all your old pictures."
	(Above are two complete but separate quotes. Note the word "cherish" is capitalized because it begins a new quote.)

With respect to American English, there is a punctuation "tall tale" that suggests using double quotation marks when quoting an entire sentence, but using single quotation marks for individual words and phrases. There is, however, no authoritative support for this practice. The only possible use for single quotation marks in American English is for a quote within a quote. For more on the use of double or single quotation marks, see *Appendix II – American English vs. British English*.

Quotation marks. There are two different styles of quotation marks: straight quotes and curly quotes. Straight quotes are also known as computer quotes or typewriter quotes. Curly quotes are commonly referred to as smart quotes or typographer's quotes.

For the purpose of written (printed) documents, we want to make sure we always use curly quotes and avoid straight quotes:

Correct	I didn't say, "I'm not game." (curly quotes)
Incorrect	I didn't say, "I'm not game." (straight quotes)

We want to avoid commingling straight quotes with curly quotes in any given document. Remember, we use "sixes" (" or ') and "nines" (' or ") for printed documents, but not straight quotes. Straight quotes find their way into a word processing document as text is copied and pasted from e-mail attachments. A good tip for getting rid of straight quotes in any word document is to use the Find/Replace feature, which is included with any word processing application.

Redundancies. Delete redundancies. Examples: Instead of writing "continued on," write "continued." Rather than writing "join together," write "join." Instead of writing "serious disaster," write "disaster." Rather than writing "tall skyscrapers," write "skyscrapers." Instead of writing "past history," write "history."

Sentence openers. Can we begin sentences with the conjunctions "and" or "but"? There is a grammar folk tale that says we shouldn't begin sentences with either of these two words, but, in fact, it is both common and accepted practice in standard written English to do so. Most writers and journalists have embraced the additional variety gained from opening sentences in this manner. It is also acceptable to begin sentences with "because." In the same way that the words "as" and "since" are often used to begin sentences, the word "because," when likewise functioning as a subordinating conjunction, may also be used to begin sentences.

Slashes. A slash (also known as a virgule) is commonly used to separate alternatives. No space should be used on either side of the slash; the slash remains "sandwiched between letters."

Incorrect	At a minimum, a résumé or CV should contain a person's job responsibilities <u>and / or</u> job accomplishments.

Correct At a minimum, a résumé or CV should contain a person's job responsibilities <u>and/or</u> job accomplishments.

Space: Break up long paragraphs. Avoid long paragraphs in succession. Break them up whenever possible. This applies to e-mails as well. Often it is best to begin an e-mail with a one- or two-sentence opener before expounding on details in subsequent paragraphs.

Space: Never two spaces after periods. Avoid placing two spaces after a period (ending a sentence). Use one space. Computers automatically "build in" proper spacing. Leaving two spaces is a carryover from the days of the typewriter.

Spacing within tables. The most common editing error when presenting tables involves adequate spacing. A practitioner's rule is to leave, within any table cell, approximately one line space above the beginning line of type and below the ending line of type. In other words, don't let the lines of the table suffocate the type. Another important thing, if using bullet points within tables, is to make sure hanging indents line up. That is, text that flows from line to another should line up below its respective bullet point.

Standard vs. nonstandard words and phrases. Because language changes over time, complete agreement never exists as to what grammatical words and phrases are considered standard. From one grammar handbook to another and from one dictionary to another, slight variations arise. These differences are due in large part to the differences between colloquial and formal written language. For example, in colloquial written English, the words "all right" and "alright" as well as "different from" and "different than" are used interchangeably. Lexicographers continue to have difficulty deciding whether to prescribe language or describe it. Should they prescribe and dictate what are the correct forms of language, or should they describe and record language as it is used by a majority of people? *Exhibit C*, on the following page, provides common misusages to watch for.

Exhibit C   Standard vs. Nonstandard Usages

Standard	Nonstandard
After all, a lot, all right	Afterall, alot, alright
Anywhere, everywhere, nowhere	Anywheres, everywheres, nowheres
Because, since, as (when these words are used as a conjunction meaning "for the reason that")	Being as/being that
Could have/would have/should have/might have/may have	Could of/would of/should of/might of/may of
Every time (always written as two words)	Everytime
Himself, themselves	Hisself, theirselves
In comparison to	In comparison with
In contrast to	In contrast with
In regard to, with regard to	In regards to, with regards to, in regards of
Regardless	Irregardless
Supposed to/used to	Suppose to/use to
The reason is that	The reason is because

Titles and capitalization. With respect to book, magazines, songs, etc., confusion often exists as to when titles are italicized and when they are placed in quotation marks. Note that underlining is no longer used to identify titles (gone are the days of the typewriter). The general rule is that longer works or full works are placed in italics. Partial works or short works are placed in quotation marks, and are not italicized. This means that the titles of books, magazines, newspapers, movies, TV programs, radio programs, plays, and names of albums are italicized. However, the titles of articles, essays, short stories, poems, chapters in a book, episodes in a TV series, and songs are placed in quotation marks.

Three rules are always observed with regard to the capitalization of titles: always capitalize the first and last words of a title and never use a period after the last word. Beyond this, the rules for capitalization of titles are somewhat arbitrary. The broad rule is to capitalize all important words and not to capitalize small, unimportant words. Important words include all nouns, pronouns, verbs, adjectives, and adverbs. Exceptions may include the verbs "is," "am," and "are" and the word "as," regardless of what part of speech it represents. "Unimportant" words—prepositions, conjunctions, and interjections—may or may not be capitalized. Two-letter prepositions (e.g., *at, by, in, of, on, to, up*) are seldom capitalized and the articles (i.e., *a, an, the*) are virtually never capitalized (unless, of course, they're the first word of a title). The coordinating conjunctions *and, but, or, nor,* and *for* are seldom capitalized; the conjunctions *yet* and *so* are almost always capitalized.

NOTE ᙏ Some confusion may arise with regard to the words "capitalization" and "full caps." Capitalization denotes placing only the first letter of a word in caps (e.g., Great). Full caps refers to placing every letter of a word in caps. (e.g., GREAT).

Weak openers. Limit the frequent use of sentences which begin with *it is, there is, there are,* and *there were.* These constructions create weak openers. A sound practice is to never begin the first sentence of a paragraph (i.e., the opening sentence) with this type of construction.

Original   <u>It is</u> obvious that dogs make better pets than hamsters.

Better   Dogs make better pets than hamsters.

Original   <u>There is</u> an excellent chance that a better diet will make you feel better.

Better   A better diet will make you feel better.

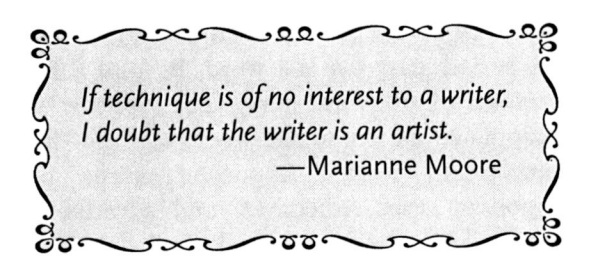

> *If technique is of no interest to a writer,
> I doubt that the writer is an artist.*
> —Marianne Moore

Editing II – Punctuation Highlights

In spoken English, we can convey our meaning through voice and body language: waving hands, rolling eyes, raising eyebrows, stress, rhythm, intonations, pauses, and even repeated sentences. In written language, we do not have such an arsenal of props; this is the unenviable job of punctuation. Mastery of punctuation, along with spelling, requires further review, and is not the focus of this book. But two key areas—commas and semicolons—are addressed because they represent areas where some of the most common punctuation errors occur.

Commas

It is said that ninety percent of writers can use the comma correctly seventy-five percent of the time, but only one percent of writers can use the comma correctly ninety-nine percent of the time. The comma is often used, but often used incorrectly. The well-known advice that a comma be used whenever there is a pause is terribly misleading. Arguably the best way to master the comma is to think of every comma as fitting into one of six categories: listing comma, joining comma, bracketing comma, contrasting comma, omission comma, or confusion comma.

Listing Comma

A listing comma separates items in a series. If more than two items are listed in a series, they should be separated by commas. The final comma in the series, the one that precedes the word *and*, is required (see *Appendix II – American English vs. British English* for further discussion about the use of a comma before a final "and").

Correct	A tostada is usually topped with a variety of ingredients, such as shredded meat or chicken, refried beans, lettuce, tomatoes, and cheese.

Do not place commas before the first element of a series or after the last element.

Incorrect	The classic investment portfolio consists, of stocks, bonds, and short-term deposits.
	Remove the comma placed after the word "consists."
Correct	The classic investment portfolio consists of stocks, bonds, and short-term deposits.

Incorrect	Conversation, champagne, and door prizes, were the highlights of our office party.
	Remove the comma placed after the word "prizes."
Correct	Conversation, champagne, and door prizes were the highlights of our office party.

Bracketing Comma

There are four main uses of the bracketing comma: (1) to set off nonessential information in the middle of a sentence; (2) to set off an opening phrase or clause; (3) to set off a closing phrase or clause; and (4) to set off speech in direct dialogue.

First, bracketing commas set off nonessential (nonrestrictive) information placed in the middle of a sentence. Such information (in the form of phrases and clauses) is not essential to the main idea of the sentence; in fact, we can test this. If after omitting words the sentence still makes sense, we know these words are nonessential and optional.

Correct	*The Tale of Genji*, written in the eleventh century, is considered by literary historians to be the world's first novel.

The main idea is that *The Tale of Genji* is considered to be the world's first novel. The intervening phrase, "written in the eleventh century," merely introduces additional but nonessential information.

Correct	The old brick house that is painted yellow is now a historical landmark.
Correct	The old brick house at O'Claire Point, which we visited last year, is now a historical landmark.

Regarding the first of two examples at the bottom of the previous page, "that is painted yellow" defines which old brick house the author is discussing. In the second example, the main point is that the old brick house at O'Claire Point is now a historical landmark, and the intervening clause "which we visited last year" merely adds additional but nonessential information.

NOTE ☙ Commas (bracketing) are not used before the second component part of a correlative conjunction (e.g., *either...or, neither...nor, not only...but also, both...and,* and *whether...or).* In the following sentence, for example, no comma should be used before "or": <u>Whether</u> you're a first-time attendee, <u>or</u> a seasoned veteran, we'll tailor our workshop to make sure you come away with valuable tips.

The second major use of the bracketing comma is to set off opening phrases and clauses from the main sentence (independent clause).

Correct	Like those of Sir Isaac Newton, the scientific contributions of Albert Einstein have proven monumental.
	A comma in the above sentence separates the prepositional phrase "like those of Sir Issac Newton" from the main sentence.
Correct	Having collected rare coins for more than fifteen years, Bill was heartbroken when his collection was stolen in a house burglary.
	A comma separates the participial phrase "having collected rare coins for more than fifteen years" from the main sentence. This participle (or participial) phrase serves as an adjective in describing Bill.

If the opening phrase is very short, the use of the comma is considered optional. In the following example, the decision whether to use a comma after "at present" rests with the writer.

Correct At present we are a crew of eight.

The third major use, though not as common as the first two uses, involves bracketing a nonessential closing phrase or clause from the main sentence (independent clause).

Correct I hope we can talk more about this idea during the conference, if time permits.

A comma is used to set off the phrase "if time permits" because this phrase functions as a piece of nonessential information. If we deleted these words, the sentence would still make sense.

Correct They woke up at 6 a.m., when they heard the rooster crowing.

Correct They woke up when they heard the rooster crowing.

The first of the above two sentences contains a nonessential clause which is bracketed. The fact that "they woke up at 6 a.m." is the critical information. The reason for their waking up is auxiliary information. However, in the second sentence, "when they heard the rooster crowing" is critical information about why they woke up. This restrictive information is not set off by commas.

NOTE ☙ A point of possible confusion occurs when a sentence ends with a phrase or clause beginning with "which." For example: "I like that new brand of coffee, which is now on sale." It is common practice to place a comma before "which" because it is assumed that such closing phrases or clauses are parenthetical. That is, they do not contain defining or essential information and should therefore be preceded by a comma. It is also common practice not to place a comma before phrases or clauses beginning with the word "that" because it is assumed that such phrases or clauses do contain defining or essential information. However, the question remains, Is a comma really necessary, especially in this short sentence? One editing trick is to substitute "that" for "which" in order to edit out the comma (along with the word "which"). Nonetheless, for those who prefer to use "which" without the comma (at least in short sentences as in the example above), one rationale for doing so is the fact that these two words—"that" and "which"—are virtually interchangeable in meaning.

The fourth major use of the bracketing comma is to set off quoted speech from the speaker.

Correct	The waitress said, "Welcome."
	"Thank you," we replied.

The same treatment is afforded to unspoken dialogue or "thought speech." Most commonly it is enclosed within quotation marks, but alternatively, it may be italicized to contrast it with actual speech.

Correct	"And what is the use of a book," thought Alice, "without pictures or conversation?"
Correct	*And what is the use of a book without pictures or conversation?*

It is not necessary to use both a speech tag (e.g., "thought Alice") and italics, since use of both techniques is redundant. Placing "thought speech" in quotation marks is common practice in nonfiction writing. Placing "thought speech" in italics is common practice in fiction writing.

Similar treatment is applied when setting off a quotation.

Correct   Was it Robert Frost who wrote, "Good fences make good neighbors"?

One important distinction arises between the direct quotations and material that is merely surrounded by quotation marks. In the latter situation, we punctuate, with reference to commas, in exactly the same manner as we would "regular" sentences. This is also the same method used for punctuating sentences when dealing with sayings, maxims, adages, aphorisms, proverbs, or mottoes.

Correct   The statement "Some cats are mammals" necessarily implies that "Some mammals are cats."

                (It's the writer's choice whether to capitalize the word "some.")

Correct   Our manager's favorite saying, "Rein in the nickels," is also his most annoying.

In the example above, commas are used because the saying "Rein in the nickels" is effectively nonessential information, the omission of which would still not destroy the sentence. Case in point: "Our manager's favorite saying is also his most annoying."

NOTE ෬ Bracketing commas are, of course, used with dates, addresses, and salutations (opening lines of letters or memos) and complimentary closes. These uses are quite common and easily understood; they are not covered here as they are unlikely to cause confusion.

Joining Comma

Use commas to separate independent clauses connected by coordinating conjunctions such as *and, but, yet, or, nor, for,* and *so.* (Independent clauses are clauses that can stand alone as complete sentences.)

Correct   Susan wants to get her story published, and she wants to have it made into a movie.

Correct   Maurice ate habanero peppers with almost every meal, yet he hardly ever got indigestion.

The following is a potentially tricky situation in which it is difficult to determine whether the comma goes before or after the *and*.

Correct   I'll put together a business plan, and by next week, I'll send it to a few potential investors.

In the previous example, there must be a joining comma before *and*, and ideally a bracketing comma after *week*. We have, after all, two complete sentences: "I'll put together a business plan" and "By next week, I'll send it to a few potential investors." Note that the comma before *and* cannot be a bracketing comma because we cannot remove the words "and by next week" without creating a run-on sentence (i.e., two sentences that are joined without proper punctuation). Note that we could put a comma after *and* (given that "by next week" is an optional phrase), but we typically do not as a matter of practice. Thus, in the next example below, the use of a third comma, although not visually pleasing, is not incorrect:

Correct   I'll put together a business plan, and, by next week, I'll send it to a few potential investors.

Correct   Some experts do not believe alcoholism should be called a disease and, moreover, believe that any type of dependency can be cured by identifying and treating its underlying causes.

In the previous example, we do not have two complete sentences, so we cannot have a joining comma (i.e., "believe that any type of dependency can be cured by identifying and treating its underlying causes" is not a complete sentence). But since the connecting word "moreover" is merely optional, it should be enclosed with commas. In other words, we could write: "Some experts do not believe alcoholism should be called a disease and believe that any type of dependency can be cured by identifying and treating its underlying causes." Knowing that we can omit a word or words and still have a sentence that makes sense is the telltale sign that we have an optional phrase.

A joining comma is optional in the case of two very short, complete sentences (independent clauses) joined by a coordinating conjunction.

Correct   The rain has stopped and the sun is shining.

Correct   The clouds are gone but it's windy.

The coordinating conjunctions "and" and "but" each join two complete sentences.

Contrasting Comma

Correct   The new music director vowed to take an active, not passive, fundraising role.

Correct   She didn't cry from sorrow and pain, but from relief and joy.

In both of the above sentences, there is sufficiently strong contrast to warrant the use of a contrasting comma.

Correct   A poorer but happier man could not be found.

In the above example, however, no commas are used to bracket the words "but happier." The important point in deciding whether to use contrasting commas rests primarily with the emphasis needed within a given sentence. Strong emphasis will require commas to separate contrasting word groups; light to moderate emphasis will not require the aid of commas. Note that the distinction regarding using or not using a pair of contrasting commas has little to nothing to do with whether the words are essential. It could be argued that all information is essential when using contrasting commas.

NOTE ⌘ When "because" joins two parts of a sentence, does a comma go before the word "because"? This is a mystery question open to debate. Consider these two examples:

Correct   Don't forget to bring an umbrella because it's going to rain out.

Correct   To tell those grief-stricken people that we know how they feel is disingenuous, because we don't know.

Some people like to place a comma before almost every use of "because." They would prefer to write, "Don't forget to bring an umbrella, because it's going to rain out." A likely better, more consistent practice is to use a comma before "because" only if it qualifies as a contrasting comma, or perhaps a bracketing comma.

Case in point: There is not a strong sense of contrast between the need to remember to bring an umbrella given the likelihood of rain. There is, however, a stronger sense of contrast in thinking we know how other grief-stricken people feel and the fact that we don't know how they feel. Most often, we will not require a comma before the word "because." First, use of the subordinating conjunction "because" creates a logical connection between ideas in a sentence, making it unlikely that the information it connects is nonessential (therefore no bracketing comma is needed). Second, the word "because" embodies a reasonable degree of contrast, obviating the need for a contrasting comma.

EDITING II – PUNCTUATION HIGHLIGHTS

Use a comma to separate word groups that flow in natural opposition to each other.

Correct   Out of sight, out of mind.

Correct   The more you practice, the better you'll get.

A contrasting comma is also used to separate two identical words in succession.

Correct   This is a great, great ice-cream flavor.

Correct   Many, many articles have been written about weight loss and weight gain.

Omission Comma

Use commas to indicate missing words. In those situations involving adjectives, the missing word is typically *and*.

Correct   I can't believe you sat through that long, dull, uninspired lecture without once checking your watch.

We can test this sentence by replacing each comma with *and*:

Correct   I can't believe you sat through that long and dull and uninspired lecture without once checking your watch.

Correct   It was a juicy, ripe mango.

Incorrect   It was a juicy ripe mango.

Incorrect   It was a juicy, ripe, mango.

A comma is required to separate *juicy* from *ripe*. There are two ways to confirm this. First, substitute *and* for the comma and see if things still make sense. (Example: "It was a juicy and ripe mango.") Second, reverse the word order and see if the sentence makes sense. (Example: "It was a ripe, juicy mango.") Either or both of these tests confirm that a comma is needed.

A comma should not be placed after *ripe* because *and* cannot be substituted for it. For instance, the phrase "ripe and mango" makes no sense. The rule is that a comma should not be placed between the modifier and the noun it modifies.

A comma can be used to take the place of omitted words.

Correct	The first playoff game was exciting; the second, dull.

In the above sentence, the comma takes the place of the "playoff game was." The sentence effectively reads: "The first playoff game was exciting; the second playoff game was dull."

Confusion Comma

A comma may be used to prevent confusion, particularly in those situations where the absence of a comma would otherwise cause the reader to misread.

Incorrect	To Karen Jane was as heroic a real-life character as could be found in any novel.
Correct	To Karen, Jane was as heroic a real-life character as could be found in any novel.

The eye cannot resist reading both names together as if they represent a first and last name.

Incorrect	Run for your life is in danger.
Correct	Run, for your life is in danger.

One could argue that a contrasting comma is needed in the above example. However, the actual problem is that the reader's eye has trouble knowing how to group the words properly.

Incorrect	The speaker said: "On Day 1 I will discuss the reasons for the global increase in diabetes and on Day 2 I will talk about how to curtail this trend."
Correct	The speaker said: "On Day 1, I will discuss the reasons for the global increase in diabetes, and on Day 2, I will talk about how to curtail this trend."

Obviously, a comma is needed in the above example to avoid confusion between the close proximity of the numbers 1 and 2 and the personal pronoun "I."

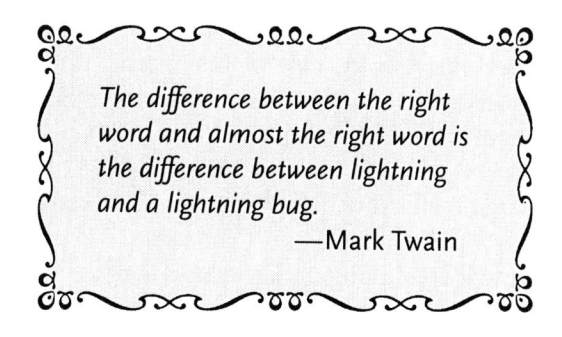

> *The difference between the right word and almost the right word is the difference between lightning and a lightning bug.*
> —Mark Twain

Exercises

Correct the comma usage in each sentence by observing its five uses: listing, bracketing, joining, contrasting, or omission.

1. The Oscar the Emmy and the Tony are three related awards which confuse many people.

2. Emerging from the ruins of the World War II Japan embarked on an economic recovery that can be only viewed in historical terms as astonishing.

3. Every major band requires, a lead singer, a lead guitarist, a bass guitarist, and a drummer.

4. A dedicated empathetic individual can achieve lifetime recognition as a United Nations worker.

5. More than a few people were shocked to discover that a torn, previously worn, pair of Madonna's underwear sold for more money at auction than did a large, splendid, sketch by Vignon.

6. The more he talked with her the more he liked her.

7. The crowded housing tenement, a cluster of rundown, look-alike apartments was the site of the Prime Minister's birthplace.

8. South Africa is famous for her gold and diamonds, Thailand, for her silk and emeralds, and Brazil for her coffee and sugarcane.

9. She reached for the clock, and finding it, hastily silenced the alarm.

10. Josie originally wanted to be a nurse but after finishing university she decided to become a flight attendant instead.

Semicolons

Use a semicolon instead of a coordinating conjunction (i.e., *and, but, yet, or, nor, for, so*) to link two closely related sentences. The key thing to remember is that a semicolon separates complete sentences. It is not used if one or more of those sentences is a fragment.

Correct   Today's students are more creative and technologically savvy, but they are also weaker in the basics of reading, writing, and arithmetic.

Correct   Today's students are more creative and technologically savvy; they are also weaker in the basics of reading, writing, and arithmetic.

Use a semicolon between independent clauses connected by words such as *however, therefore, moreover, nevertheless,* and *consequently*. These special words are called conjunctive adverbs.

Incorrect   The formulas for many scientific discoveries appear rudimentary, however, when one examines a derivation behind these formulas they do not seem so rudimentary after all.

Correct   The formulas for many scientific discoveries appear rudimentary; however, when one examines a derivation behind these formulas they do not seem so rudimentary after all.

Run-on Sentence

To see commas and semicolons in action, let's review a very common error—the run-on sentence. A run-on refers to two sentences that are inappropriately joined together, usually by a comma. There are effectively four ways to correct a run-on sentence, as seen in each of the four correct options below. First, join the two sentences with a semicolon. Second, join the two sentences with a coordinating conjunction (e.g., *and, but, yet, or, nor, for, so*). Third, separate the two sentences with a period. Fourth, turn one of the two sentences into a subordinate clause.

Incorrect	Technology has made our lives easier, it has also made our lives more complicated.
Correct	Technology has made our lives easier; it has also made our lives more complicated.
	This solution involves changing the comma to a semicolon.
Correct	Technology has made our lives easier, and it has also made our lives more complicated.
	This solution involves joining two sentences with a coordinating conjunction.
Correct	Technology has made our lives easier. It has also made our lives more complicated.
	This solution involves making two separate sentences.
Correct	Even though technology has made our lives easier, it has also made our lives more complicated.
	This solution involves turning one sentence into a subordinate clause. Here the focus is on the idea that technology has made our lives more complicated (independent clause). The subordinate idea is that it has made our lives easier (subordinate clause). It would also be equally correct to say: "Even though technology has made our lives more complicated, it has also made our lives easier." Now the central idea and subordinate idea are reversed.

Answers to Exercises

1. The Oscar, the Emmy, and the Tony are three related awards which confuse many people.

   The comma after Emmy is required in American English but omitted in British English.

2. Emerging from the ruins of the World War II, Japan embarked on an economic recovery that can only be viewed in historical terms as astonishing.

   A bracketing comma is required after "World War II."

3. Every major band requires a lead singer, a lead guitarist, a bass guitarist, and a drummer.

   There should be no comma after the verb "requires."

4. A dedicated, empathetic individual can achieve lifetime recognition as a United Nations worker.

   An omission comma separates "dedicated and empathetic." There are two ways to test for this. First, substitute the word "and" to read "dedicated and empathetic." Second, reverse the order of the two words to read "empathetic, dedicated individual." Since either substituting the word "and" or reversing the word order still makes sense in context, a comma should be used.

5. More than a few people were shocked to discover that a torn, previously worn pair of Madonna's underwear sold for more money at the auction than did a large, splendid sketch by Vignon.

   There are no commas after "previously worn" or "sketch." A comma is not placed between a modifier and the word it modifies. Here the words being modified are a "pair of Madonna's underwear" and "sketch."

6. The more he talked with her, the more he liked her.

   A contrasting comma after "her" is required.

7. That crowded housing tenement, a cluster of run-down, look-alike apartments, was the site of the Prime Minister's birthplace.

   Insert a comma after "apartments"; the phrase "a cluster of run-down, look-alike apartments" is nonessential (and therefore optional) and should be enclosed with commas.

8. South Africa is famous for her gold and diamonds, Thailand, for her silk and emeralds, and Brazil, for her coffee and sugarcane.

   A comma is needed after Thailand and Brazil. Such a comma (an omission comma) takes the place of the words "is famous." So, the sentence effectively reads: "South Africa is famous for her gold and diamonds, Thailand is famous for her silk and emeralds, and Brazil is famous for her coffee and sugarcane."

   There are at least two additional ways to correct this sentence:

   i) By omitting the second comma in the original:

   "South Africa is famous for her gold and diamonds, Thailand for her silk and emeralds, and Brazil for her coffee and sugarcane."

   The treatment is consistent with the rules of ellipsis. We can acceptably omit words (in this case the words "is famous") when they are readily understood in context.

ii) By using semicolons with commas:

"South Africa is famous for her gold and diamonds; Thailand, for her silk and emeralds; and Brazil, for her coffee and sugarcane."

Semicolons can be used in conjunction with commas, especially in cases of heavily punctuated sentences. The final "and" appearing before "Brazil" is optional.

9. She reached for the clock and, finding it, hastily silenced the alarm.

A bracketing comma is needed before and after the words "finding it"; this is a nonessential phrase. Removing these words still results in a complete sentence. Case in point: "She reached for the clock and hastily silenced the alarm." If, however, we were to remove the words "and finding it," the sentence would become nonsensical: "She reached for the clock hastily silenced the alarm." Therefore this confirms that a set of bracketing commas cannot be used in the original sentence.

10. Josie originally wanted to be a nurse, but after finishing university, she decided to become a flight attendant instead.

A joining comma is required before "but," while a bracketing comma is required after "university." We effectively have two sentences: "Josie originally wanted to be a nurse" and "After finishing university, she decided to become a flight attendant." In the solution above, the two commas do not both function as bracketing commas; if this were so we could cut out the phrase "but after finishing university" and the sentence would still make sense, but it doesn't: "Josie originally wanted to be a nurse she decided to become a flight attendant instead."

American English vs. British English

American English and British English are the two major engines behind the evolving English language. Other English-speaking countries—most notably Canada, Australia, New Zealand, India, the Philippines, and South Africa—embrace a variant of one or both of these two major systems. Although American and British English do not differ with respect to grammar per se, each system has its own peculiarities in terms of spelling and punctuation. The purpose of this section is to provide a snapshot of these differences.

Spelling Differences

Spelling Differences Between American English and British English

American English		British English	
-ck	check	-que	cheque
-ed	learned	-t	learnt
-er	center, meter	-re	centre, metre
-no e	judgment, acknowledgment	-e	judgement, acknowledgement
-no st	among, amid	-st	amongst, amidst
-in	inquiry	-en	enquiry
-k	disk	-c	disc
-l	traveled, traveling	-ll	travelled, travelling
-ll	enroll, fulfillment	-l	enrol, fulfilment
-m	program	-mme	programme
-o	mold, smolder	-ou	mould, smoulder
-og	catalog	-ogue	catalogue
-or	color, favor	-our	colour, favour
-s	defense, offense	-c	defence, offence
-z	summarize, organization	-s	summarise, organisation

Spelling fine points. The British generally double the final *-l* when adding suffices that begin with a vowel, where Americans double it only on stressed syllables. This makes sense given that American English treats *-l* the same as other final consonants, whereas British English treats it as an exception. For example, whereas Americans spell *counselor, equaling, modeling, quarreled, signaling, traveled,* and *tranquility,* the British spell *counsellor, equalling, modelling, quarrelled, signalling, travelled,* and *tranquillity.*

Certain words—*compelled, excelling, propelled,* and *rebelling*—are spelled the same on both platforms, consistent with the fact that the British double the *-l* while Americans observe the stress on the second syllable. The British also use a single *-l* before suffixes beginning with a consonant, whereas Americans use a double *-l.* Thus, the British spell *enrolment, fulfilment, instalment,* and *skilful,* Americans spell *enrollment, fulfillment, installment,* and *skillful.*

Deciding which nouns and verbs end in *-ce* or *-se* is understandably confusing. In general, nouns in British English are spelled *-ce* (e.g., *defence, offence, pretence)* while nouns in American English are spelled *-se* (e.g., *defense, offense, pretense).* Moreover, American and British English retain the noun-verb distinction in which the noun is spelled with *-ce* and the core verb is spelled with an *-se.* Examples include: *advice* (noun), *advise* (verb), *advising* (verb) and *device* (noun), *devise* (verb), *devising* (verb).

With respect to *licence* and *practice,* the British uphold the noun-verb distinction for both words: *licence* (noun), *license* (verb), *licensing* (verb) and *practice* (noun), *practise* (verb), *practising* (verb). Americans, however, spell *license* with a *-s* across the board: *license* (noun), *license* (verb), *licensing* (verb), although *licence* is an accepted variant spelling for the noun form. Americans further spell *practice* with a *-c* on all accounts: *practice* (noun), *practice* (verb), *practicing* (verb).

Punctuation Differences

The following serves to highlight some of major differences in punctuation between America English and British English.

Abbreviations

American English	*British English*
Mr. / Mrs. / Ms.	Mr / Mrs / Ms

Americans use a period (full stop) after salutations; the British do not.

American English	*British English*
Nadal vs. Federer	Nadal v. Federer

Americans use "vs." for versus; the British write "v." for versus. Note that Americans also use the abbreviation v. in legal contexts. For example, Gideon v. Wainright.

Colons

American English	*British English*
We found the place easily: Your directions were perfect.	We found the place easily: your directions were perfect.

Americans often capitalize the first word after a colon, if what follows is a complete sentence. The British prefer not to capitalize the first word that follows the colon, even if what follows is a full sentence.

Commas

American English	British English
She likes the sun, sand, and sea.	She likes the sun, sand and sea.

Americans use a comma before the "and" when listing a series of items. The British do not use a comma before the "and" when listing a series of items.

American English	British English
In contact sports (e.g., American football and rugby) physical strength and weight are of obvious advantage.	In contact sports (e.g. American football and rugby) physical strength and weight are of obvious advantage.

The abbreviation "i.e." stands for "that is"; the abbreviation "e.g." stands for "for example." In American English, a comma always follows the second period in each abbreviation (when the abbreviation is used in context). In British English, a comma is never used after the second period in either abbreviation.

Note that under both systems, these abbreviations are constructed with two periods, one after each letter. The following variant forms are *not* correct under either system: "eg.," or "eg." or "ie.," or "ie."

Dashes

American English	British English
The University of Bologna – the oldest university in the Western World – awarded its first degree in 1088.	The University of Bologna - the oldest university in the Western World - awarded its first degree in 1088.

The British have traditionally favored the use of a hyphen where Americans have favored the use of the dash. Discussion of the two types of dashes is found in *Editing I – Tune-up*.

Quotation Marks

American English	British English
Some see education as a "vessel to be filled," others see it as a "fire to be lit."	Some see education as a 'vessel to be filled', others see it as a 'fire to be lit'.

Americans use double quotation marks. The British typically use single quotation marks.

American English	British English
Our boss said, "The customer is never wrong." Or: "The customer is never wrong," our boss said.	Our boss said, 'The customer is never wrong.' Or: 'The customer is never wrong,' our boss said.

Periods and commas are placed inside quotation marks in American English (almost without exception). In British English, the treatment is twofold. Punctuation goes inside quotation marks if it's part of the quote itself; if not, quotation marks go on the outside. This means that in British English periods and commas go on the outside of quotation marks in all situations not involving dialogue or direct speech. However, in situations involving direct speech, periods and commas generally go inside of quotation marks because they are deemed to be part of the dialogue itself.

NOTE ☙ Today, the practice of using single quotation marks is not ubiquitous in the United Kingdom. A number of UK-based newspapers, publishers, and media companies now follow the practice of using double quotation marks.

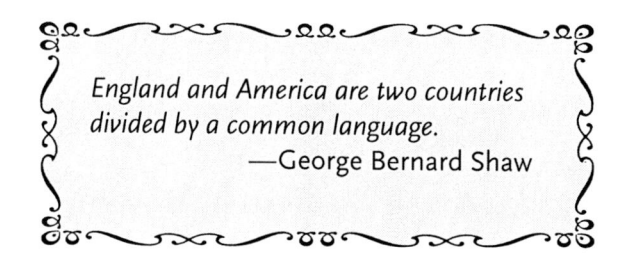

> England and America are two countries divided by a common language.
> —George Bernard Shaw

Traditional Writing vs. Digital Writing

There are two different platforms across which writing takes place—traditional, paper-based and electronic, digital-based. Three questions may arise when writing for these two different media: Should writing across different platforms be different, and if so, how is it different and what dynamics cause this difference? The debate over whether there should be a difference in terms of writing standards is essentially a values debate to which there is no correct answer. However, from a practical standpoint, there is little doubt that written communication across these media differs and will remain different.

The dynamics that cause a difference between traditional and digital writing center on the distinction between static and non-static written communication. Electronic communication arguably exists to take the place of spoken communication, and to that extent, it is non-static, tending to be more conversational and less structured.

Traditional or paper-based writing is most often associated with formal writing, while electronic or digital-based writing is commonly the domain of informal writing. As mentioned, traditional writing overlaps in large degree with formal writing. Formal writing, loosely defined, is writing consisting of multiple paragraphs that is meant to be distributed and read by one or more persons. Examples of formal written documents include long e-mails, letters, newsletters, news articles, brochures, essays, reports, manuals, and books.

Digital writing dovetails with informal writing. Examples of digital writing include short personal and business e-mails, text messaging, blogging, and messaging on social network sites such as MySpace, Facebook, and Twitter.

How do the different forms of written communication stack up in terms of their likely level of written formality? As briefly summarized next, the higher the technological level, the more informal writing tends to be. Naturally, this analysis embodies a degree of generality. Blogging, for example, can be quite formal, as is the case with blog articles published by the Huffington Post. Most blog responses, however, tend to be as informal as are casual e-mails or text messaging.

Writing—Levels of Informality

Level	Examples
Fifth level	text messaging, instant messaging (e.g., Yahoo, Skype), and microblogging (Twitter)

Fourth level   blogging, e-mails

Third level   letter writing, articles, newsletters, brochures, websites, memos, flyers, slide shows

Second level   manuals, business reports, academic essays

First level   published documents such as reports, newspapers, magazines, and books

Some of the telltale signs of informality in the digital realm (for better or worse) include the following: shorter sentences; optional punctuation, including non-capitalized words and abbreviated spelling; the use of fewer adjectives and adverbs; the frequent use of ellipses, asterisks, and exclamation points; and the occasional use of smileys and e-mail acronyms.

Many traditionalists object to the use of abbreviated spelling and non-capitalization. For example, dashing off "c u @12 - lunch" translates as "see you at 12 o'clock for lunch." But consider the level of informality. Assuming this to be a text message, the message, once sent and received, will never be seen again. So what purpose would it serve to make it more formal?

Such informality in written communication may be acceptable for text messaging, but it is considered unacceptable when used in standard expository writing. The point here is that although there is real logic as to why writing may be informal, this does not mean that informal writing is superior to formal writing because it is more practical, less structured, or somehow more authentic. Formal and informal writing are different and serve different purposes.

Also, many readers understand only the most basic smileys and e-mail acronyms. Certain e-mail acronyms are easy to understand—FYI ("For Your Information"), IMHO ("In My Humble/Honest Opinion"), and LOL ("Laughing Out Loud"). But where is the line of readability to be drawn? The average e-mail user would have a devil of a time deciphering any of the following: FYEO (For Your Eyes Only), PMFJI (Pardon Me For Jumping In), and IITYWYBMAD? (If I Tell You Will You Buy Me A Drink?). The same situation holds true for smileys (emoticons), for which actual Smiley Dictionaries exist. Smileys are read by turning the head counterclockwise and looking at them sideways; then the little faces can by seen. Most readers understand :-), ;-), and :-(to mean "I'm happy/it's funny," "winking/I think I'm being funny," and "I'm sad/it's sad." But other smileys are enigmatic for the uninitiated, notably: ;-\ (undecided), :-< (very upset), and :-# (my lips are sealed).

Certain writing techniques used in digital communications occur because typographical tools are limited or unavailable. Basic e-mail, for instance, doesn't provide a way to italicize or underline. And underlining of digital text should be avoided, as it is reserved for use as hyperlinks. In order to place emphasis on certain words and phrases, it is common practice to place them in asterisks or, occasionally, to capitalize them. An employee who e-mails, "You won't believe how *smoothly* our morning meeting went," is drawing attention to the word "smoothly" for the purpose of infusing a little sarcasm, as if things hadn't gone so smoothly after all.

The way information is read on the Internet influences how it is written. Individuals don't read information online as linearly as they do in printed formats. They tend to skip around, skimming and scanning, then stopping to read chunks of information. How does this affect the way information is written for the web? Columns tend to be narrower (usually not more than seventy-five characters per line), sentences and paragraphs tend to be shorter, more heads and subheads are used to assist the reader in "grabbing" information, and more bolding is commonly used (both in black and in color).

Benefits

In conclusion, the higher the standard one adheres to in all written communication, be it digital or print, the higher will be one's perceived level of professionalism. The cost is time and effort; the benefit is quality. Each individual must make his or her own "call." As the English language continues to evolve, with the digital revolution playing a significant role in this evolution, there will always exist a place for "good" writing. Writing that is strong in content, and equally adheres to currently accepted principles and rules—including grammar, spelling, and punctuation—will continue to have a positive influence on its readers.

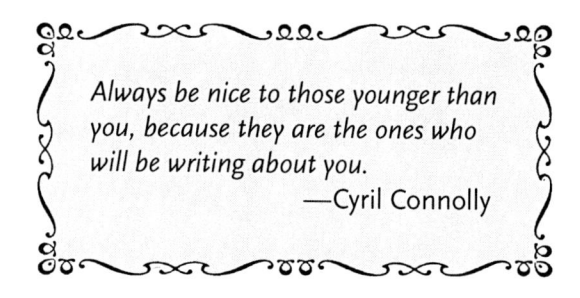

> *Always be nice to those younger than you, because they are the ones who will be writing about you.*
> —Cyril Connolly

Selected Bibliography

The Chicago Manual of Style. 15th ed. Chicago: University of Chicago Press, 2003.

Cook, Claire Kehrwald. *Line by Line: How to Edit Your Own Writing.* Boston: Houghton Mifflin, 1985.

Ehrenhaft, George. *Barron's SAT Writing Workbook.* 2nd ed, Barron's Educational Series. Hauppage, NY: Barron's, 2009.

Encarta Webster's Dictionary of the English Language. 2nd ed. New York: Bloomsbury, 2004.

Fogarty, Mignon. "Grammar Girl: Quick and Dirty Tips for Better Writing." http://grammar.quickanddirtytips.com/.

Fogiel, Max. *The English Handbook of Grammar, Style, and Composition.* Piscataway, NJ: Research and Education Association, 1987.

Kramer, Melinda, Glenn H. Leggett, and C. David Mead. *Prentice Hall Handbook for Writers.* 11th ed. Englewood Cliffs, NJ: Prentice Hall, 1995.

Merriam Webster's Collegiate Dictionary. 11th ed. Springfield, MA: Merriam Webster, 2005.

Opdycke, John B. *Harper's English Grammar.* New York: Warner, 1983.

Oxford Dictionary of English. 2nd ed. New York: Oxford University Press, 2005.

Ritter, Robert M. *The Oxford Style Manual.* New York: Oxford University Press, 2003.

Skillin, Marjorie E., and Robert M. Gay. *Words into Type.* 3rd ed. Englewood Cliffs, NJ: Prentice-Hall, 1974.

Strunk, William, Jr., and E. B. White. *The Elements of Style.* 4th ed. New York: Allyn and Bacon, 2000.

Trask, Robert Lawrence. *The Penguin Guide to Punctuation.* London: Penguin, 1999.

Truss, Lynne. *Eats, Shoots & Leaves: The Zero Tolerance Approach to Punctuation.* New York: Gotham, 2004.

Venolla, Jan. *Write Right! A Desktop Digest of Punctuation, Grammar, and Style.* 4th ed. Berkeley: Ten Speed Press, 2004.

Warriner, John E. *English Composition and Grammar: Complete Course.* Orlando, FL: Harcourt Brace Jovanovich, 1988.

Wikipedia, The Free Encyclopedia, s.v. "Writing Style," http://en.wikipedia.org/wiki/Writing_style.

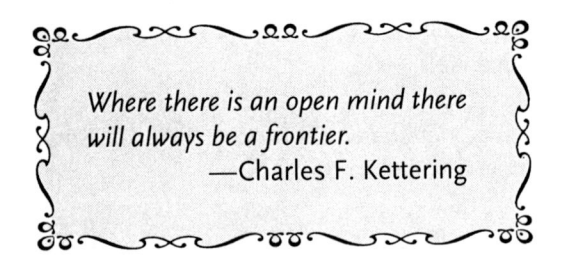

> Where there is an open mind there will always be a frontier.
> —Charles F. Kettering

Index

Numbers in *italics* (within brackets) indicate multiple-choice problem numbers, 1 to 30. They are preceded by corresponding page numbers.

active vs. passive voice
    defined, 59–60
    examples in editing, 166–67
    problem, 106–7 *(8)*

adjective, defined, 56

adjective clause, defined, 61

adverb, defined, 56

American vs. British English,
    punctuation differences, 198–200
    spelling differences, 196–97

antecedent, defined, 61

appositive phrase, defined, 61

article, defined, 61

blogs (blogging), writing for, 204–5

British English. *See* American vs. British English

case, defined, 59

clause, defined, 61–62

collective noun, defined, 62

comma
    bracketing, 177–81
    confusion, 186–87
    contrasting, 183–85
    joining, 181–83
    listing, 176–77
    omission, 185–86, 111 *(15)*

comparisons
    "apples with apples," the logic of comparing two things, 22, 115–16 *(23–24)*
    comparative vs. superlative, 22, 111 *(16)*
    demonstrative pronouns, "those" and "that," 23, 112 *(17–18)*
    idioms, situations involving, 113–15 *(20–22)*
    "like" vs. "as," 24, 113 *(19)*
    problems, multiple choice, 111–16 *(16–24)*

complement, defined, 62

conjunction
    beginning sentences with "and" or "but," 171
    coordinating conjunction, defined, 62
    correlative conjunction, defined, 62
    problems, 102 *(1)*, 110 *(14)*
    using without commas, 178

coordinating conjunction. *See* conjunction

correlative conjunction. *See* conjunction

demonstrative pronoun, defined, 63

dependent clause, defined, 63

diction
   affect, effect, 74
   afterward, afterwards, 74
   allot, alot, a lot, 74
   all ready, already, 74–75
   all together, altogether, 75
   among, amongst, 75
   anymore, any more, 76
   anyone, any one, 76
   anytime, any time, 76
   anyway, any way, 76–77
   apart, a part, 77
   awhile, a while, 77
   as, because, since, 77
   assure, ensure, insure, 78
   because of, due to, owing to, 78
   better, best, 78
   between, among, 79
   cannot, can not, 79
   choose, choosing, chose, chosen, 79
   complement, compliment, 80
   complementary, complimentary, 80
   differs from, differ with, 80–81

(*continued*)
   different from, different than, 81
   do to, due to, 81
   each other, one another, 81
   everyday, every day, 81–82
   everyplace, every place, 82
   everyone, every one, 28, 82
   everything, every thing, 82
   every time, everytime, 83
   farther, further, 83
   fewer, less, 83
   if, whether, 83–84
   instead of, rather than, 84
   infer, imply, 84
   into, in to, 29, 84
   its, it's, 85
   lead, led, 85
   lets, let's, 28–29, 83–84
   lie, lay, 86
   like, such as, 87
   loose, lose, loss, 87
   maybe, may be, 87
   might, may, 88
   no one, noone, 88
   number, amount, 88
   onto, on to, 88–89
   passed, past, 89
   principal, principle, 89
   sometime, some time, 89
   than, then, 90
   that, which, 28, 90
   that, which, who, 90–91
   there, their, they're, 91
   toward, towards, 91
   used to, use to, 29–30, 91
   who vs. whom, 15, 92–93
   whose vs. who's, 93
   your vs. you're, 93

digital writing
    versus traditional writing, 203–6
    formality vs. informality, 204–5

direct object, defined, 63

editing
    "a" vs. "an," 150
    abbreviations, Latin, 150–52
    apostrophes, in place of omitted letters, 152
    brevity, 152–53
    bullets vs. hyphens or asterisks, 153–54
    bulleted lists, proper form, 154–57
    colon, 157–58
    compound adjectives, 158–61
    dashes, en vs. em, 161–62
    editing on screen vs. hard copy, 168
    hyphens, 162–63
    nominalizations, 163–64
    numbers, written numbers or spelled out, 164
    page numbering, 164–65
    paragraph style, block-paragraph format vs. indented paragraph format, 165–66
    passive voice vs. active voice, 59–60, 126, 166–67
    possessives, how to form, 167–68
    qualifiers, cleaning out, 168–69
    quotations, how to punctuate, 169

(*continued*)
    quotation marks, straight quotes vs. curly quotes, 170
    redundancies, 170
    sentence openers with "and," "but," and "because," 170
    slashes, 173–74
    space
        breaking up long paragraphs, 171
        two spaces after periods, omitting, 171
        within tables, 171
    standard vs. nonstandard words and phrases, 171–72
    titles, punctuation and capitalization of, 173
    weak openers, avoiding, 173–74

ellipsis, 21, 110 *(13)*, 111 *(15)*

emails, writing of, 204–6

fragment. *See* sentence fragment

gender, defined, 58

gerund, defined, 63–64

grammar
    categories of, "big six"
        comparisons, 22–23, 111–16 *(16–24)*
        modification, 17–19, 108–11 *(10–15)*
        parallelism, 19–21, 110–11 *(13–15)*
        pronoun usage, 13–17, 108–9 *(6–9)*

213

subject-verb agreement,
10–13, 102–4 *(1–5)*
102–105 *(2–7)*
verb tenses, 24–27, 116–20 *(25–30)*

idioms
list of, 94–99
review of common, 30–33
problems, multiple-choice, 102 *(1)*, 106 *(7)*, 113–14 *(20–21)*

indefinite pronoun
chart of, 10
defined, 64

independent clause, defined, 64

indirect object, defined, 64–65

infinitive, defined, 65

interjection, defined, 57

Internet. See digital writing

interrogative pronoun, defined, 65

intransitive verb, defined, 65

modification
modifiers
dangling, 18
misplaced, 17, 108 *(10)*
squinting, 18
problems, multiple choice, 108–9 *(10–12)*

mood
defined, 60
subjunctive, "was" vs. "were," 26–27

nonrestrictive clause, defined, 65–66

nonstandard words and phrases, 171–72

noun, defined, 56

number, defined, 58

object, defined, 66

100-question quiz, the, 8–54

parallelism
correlative conjunctions, 20, 102 *(1)*, 110 *(14)*
ellipsis (omitted words), 21, 110 *(13)*, 111 *(15)*
gerunds and infinitives, 21, 104 *(5)*
problems, multiple choice, 110–11 *(13–15)*
series of three (or more) items, 20, 111 *(15)*
verbs, with respect to, 19–20

parenthetical expression, defined, 66–67

participle, defined, 67

participle phrase
defined, 67
problem, 109 *(12)*

passive voice
   defined, 59–60
   examples in editing, 166–67
   problems, 106–7 *(8)*, 119 *(28)*

person, defined, 58–59

personal pronoun
   chart of, 13
   defined, 67–68
   problems, 105–6 *(6–7)*

phrase, defined, 68

predicate, defined, 68

preposition, defined, 57

prepositional phrase. *See* subject-verb agreement

pronoun, defined, 56

pronoun usage
   pronoun-antecedent
      agreement, 15–16, 106–7
      (8)
   problems, multiple choice,
      105–7 *(6–9)*
   pronoun reference,
      ambiguous, 16, 105–6
      (6–7)
   pronoun shifts (person or
      number), 16–17, 107 *(9)*
   reflexive pronouns, 15
   subjective vs. objective forms
      subjects vs. objects of
         verbs, 15
      prepositions, direct objects
         of, 14

(continued)
   comparisons using "than"
      or "as...as," 14
   "who" vs. "whom," 15

reflexive pronoun, defined, 68

relative clause, defined, 69

relative pronoun, defined, 69

restrictive clause, defined, 69

run-on sentence, defined, 70,
   111 *(15)*, 189–90

semi-colon, 189, 111 *(15)*

sentence, defined, 70

sentence fragment, defined,
   70–71, 111 *(15)*

split infinitive, defined, 71

subject-verb agreement
   collective nouns, 12
   correlative conjunctions, 11,
      102 *(1)*
   gerunds and infinitives, 10,
      104 *(5)*
   indefinite pronouns, 11–12
   percents and fractions, 12
   prepositional phrases, 19,
      102–3 *(1–2)*
   problems, multiple choice,
      102–4 *(1–5)*
   pseudo compound subjects, 8

(*continued*)
    measurements involving money, time, weight, or volume, 12–13
    subjects joined by "and," 8
    "the number" vs. "a number," 12–13
    verbs before subjects, situations involving "here is," "here are," "there is," and "there are," 9–10, 104 *(4)*

subjunctive mood, "was" vs. "were," 26–27, 60

subordinate clause
    correcting run-on sentences, 189–90
    defined, 71

subordinating conjunction
    beginning sentences with "becuase," 171
    defined, 72

tense
    chart of, 24
    defined, 60
    See also verb tenses

text messaging, 204–5

transitive verb, defined, 72

twittering (tweaks), 204

verb, defined, 56

verb tenses
    active vs. passive voice, 59–60, 126, 166–67
    conditional verb form, "will" vs. "would," 27
    consistent use of, 26
    simple vs. progressive verb forms, chart of, 24
    future perfect tense, 25, 120 *(30)*
    past perfect tense, 35, 44–49, 117 *(26)*, 119 *(28)*
    perfect tenses, understanding differences, 26
    present perfect tense, 26, 116 *(25)*
    problems, multiple-choice, 116–20 *(25–30)*
    future tense, simple, 25, 119–20 *(29)*
    past tense, simple, 26, 118 *(27)*
    present tense, simple, 26
    See also mood; subjunctive mood

verbal, defined, 72

voice, defined, 59–60

web-based writing. *See* digital writing

who vs. whom, 15, 92–93

About the Author

Brandon Royal is an award-winning author whose educational authorship includes *The Little Red Writing Book*, *The Little Gold Grammar Book*, *The Little Green Math Book*, and *The Little Blue Reasoning Book*. During his tenure working in Hong Kong for US-based Kaplan Educational Centers—a Washington Post subsidiary and the largest test-preparation organization in the world—Brandon honed his theories of teaching and education and developed a set of key learning "principles" to help define the basics of writing, grammar, math, and reasoning. A Canadian by birth and graduate of the University of Chicago's Booth School of Business, his interest in writing began after completing writing courses at Harvard University. Since then he has authored ten books and reviews of his books have appeared in Time Asia magazine, Publishers Weekly, Library Journal of America, Midwest Book Review, The Asian Review of Books, Choice Reviews Online, Asia Times Online, and About.com. Brandon is a two-time winner of the International Book Awards, a four-time gold medalist at the President's Book Awards, as well as winner of the Global eBook Awards, the USA Book News "Best Book Awards," and recipient of the 2011 "Educational Book of the Year" award as presented by the Book Publishers Association of Alberta.

To contact the author:
E-mail: contact@brandonroyal.com
Web site: www.brandonroyal.com

Books by Brandon Royal

The Little Red Writing Book:
20 Powerful Principles of Structure, Style and Readability

The Little Gold Grammar Book:
Mastering the Rules That Unlock the Power of Writing

The Little Green Math Book:
30 Powerful Principles for Building Math and Numeracy Skills

The Little Blue Reasoning Book:
50 Powerful Principles for Clear and Effective Thinking

Secrets to Getting into Business School:
100 Proven Admissions Strategies to Get You Accepted at the MBA Program of Your Dreams

Chili Hot GMAT:
200 All-Star Problems to Get You a High Score on Your GMAT Exam

Chili Hot GMAT Math Review

Chili Hot GMAT Verbal Review

Bars of Steel:
Life before Love in a Hong Kong Go-Go Bar – The True Story of Maria de la Torre

Pleasure Island:
You've Found Paradise, Now What? A Modern Fable on How to Keep Your Dreams Alive